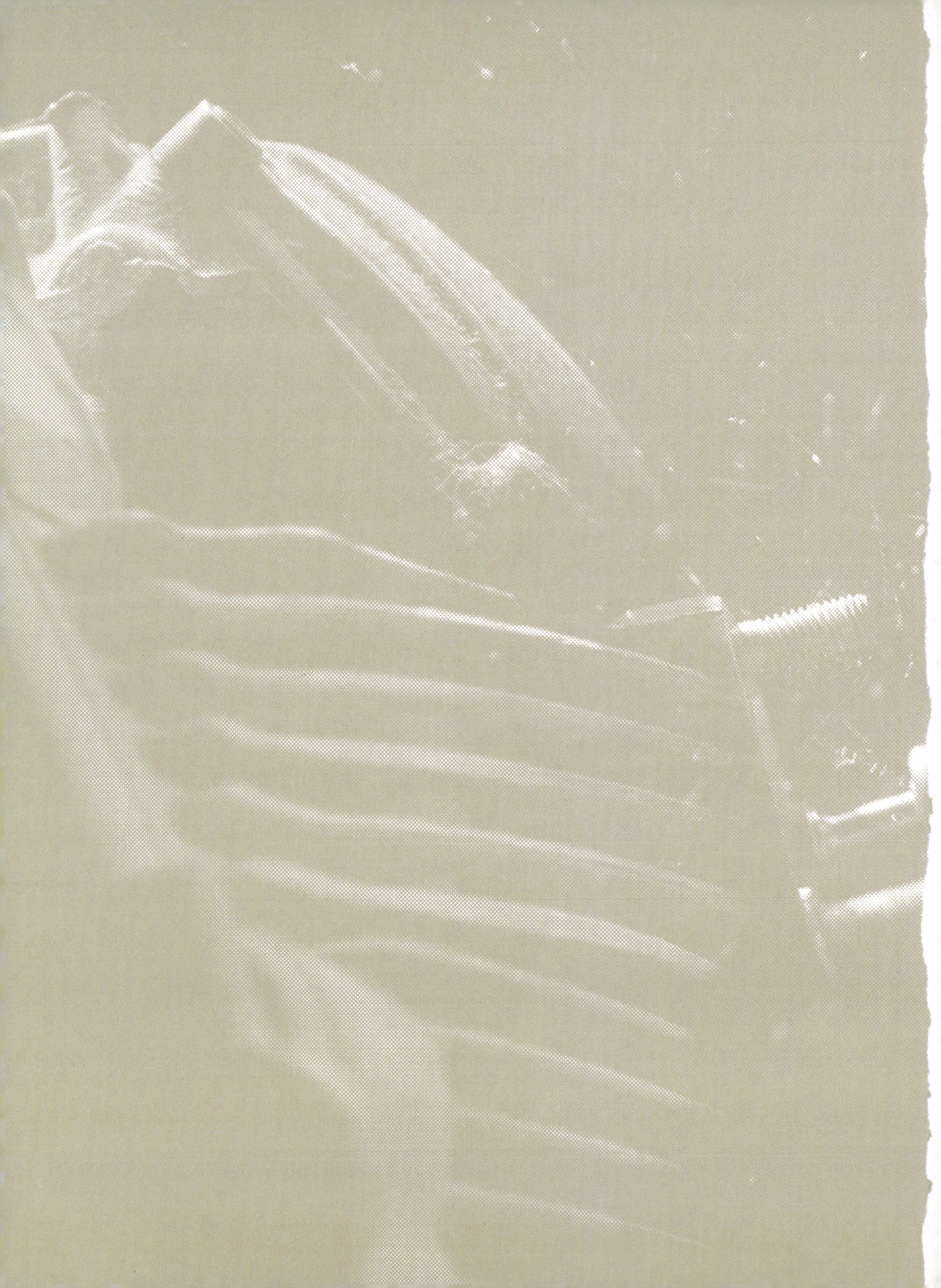

HONDA

Octane Press, Edition 1.0, June 2016

ISBN 978-1-937747-57-2

Cover and back cover photographs by Adam Ewing
Title Page photograph by Lee Klancher
Contents page photograph by David Hans Cooke

Copyedited by Dennis Pernu
Proofread by Jill Amack
Cover and Interior Design by Geoffrey McCarthy

octanepress.com | 512-334-9441

Printed in China

THE BUILD

HOW THE MASTERS DESIGN CUSTOM MOTORCYCLES

ROBERT HOEKMAN JR

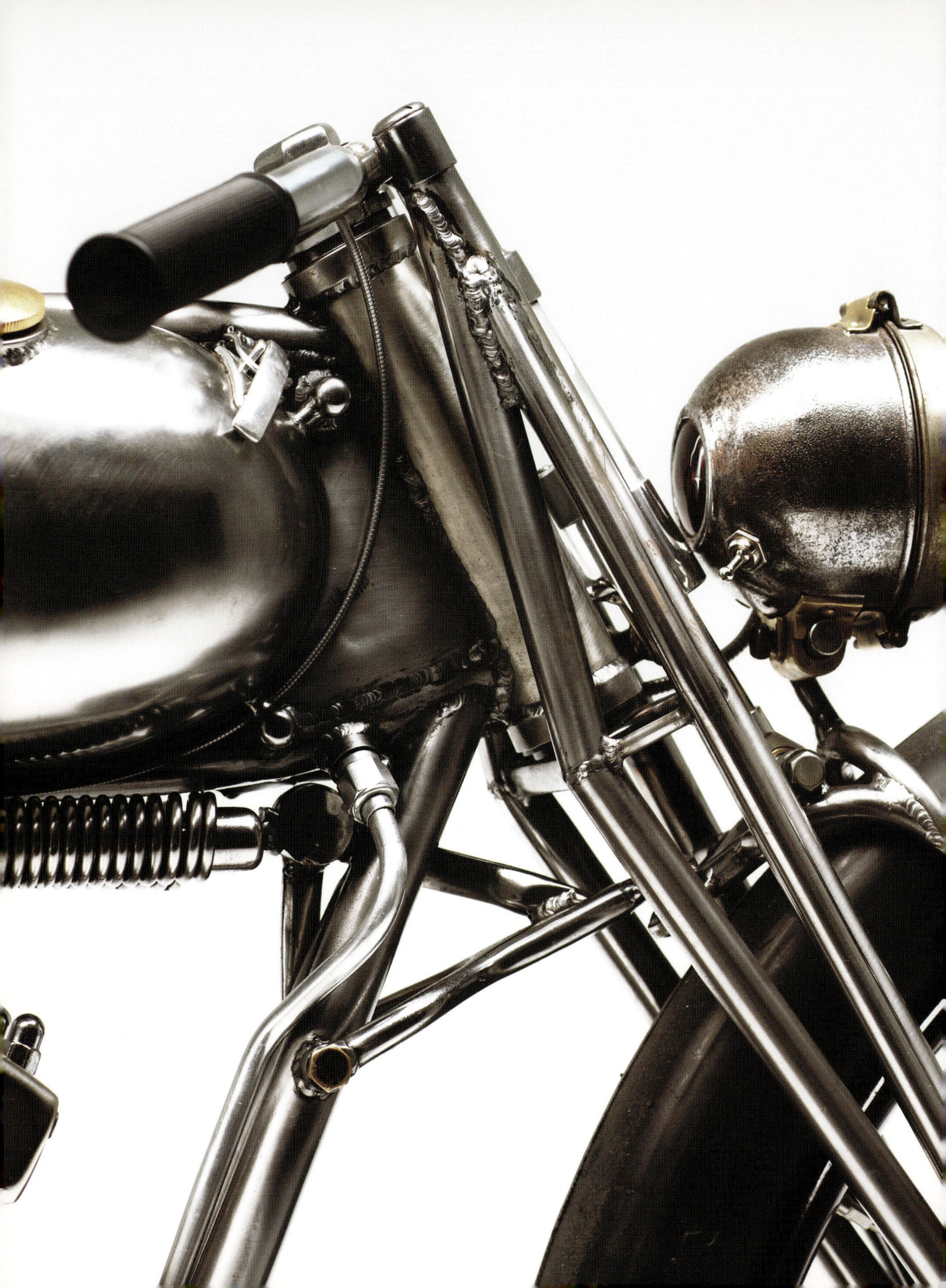

CONTENTS

CRAFTSMAN

INTRODUCTION

You know the urge.

It's found you daydreaming more than once.

That urge to build. To define. To bend, shape, fabricate, invent, shove, break. To slide your leg over the seat you finally got back from the leather shop. To twist back the throttle grip you wrapped yourself. To lunge into the darkness of an open highway, engine screaming, wind whipping through you as you ride a creation all your own. To tell the tale of the build over breakfast before a Saturday ride to the hill country.

previous spread:
David K. Browning / E3

following page:
The team at Revival spending the day haunting acclaimed builder Shinya Kimura's backyard.
Revival Cycles

You know it's about more than a motorcycle. It's about your identity. It's about finding—making—something that's an extension of yourself. Something to represent you. Something as unique as you are.

Yeah, you know the urge.

It's time you did something about it.

It's time to reply to that Craigslist ad you've been eyeing. It's time to borrow your dad's truck so you can haul home the junked-out CB550 you find at the other end. It's time to pull out the wrenches and get to work on the bike no one can build but you. A design inspired by a hundred others but conjured up by you and you alone.

It's time to find *your* bike. *Your* style. *Your* voice.

You may not yet have any sense of what that means. Writers and artists and musicians spend lifetimes looking for it, torture themselves in pursuit of it. Some create great things along the way. Some just feel it out one decision at a time and manage to crush down the self-doubt long enough to see where the decisions take them.

The iconic motorcycle designers featured in this book have all done just that. Some are master craftsmen. Some can barely draw a straight line with a ruler. They're all masters of their style.

For the first time, they've come together to help you find the one you'll call your own.

STYLES, TRENDS, AND OUTCASTS

If they are to be considered outliers of the custom motorcycle scene, the builders featured here have hardly noticed. They can't even see the scene from their shop windows. The consensus among the five gentlemen whose voices you'll hear throughout this inspiration manual, and throughout your own build, is that they simply don't have time to notice.

"Our involvement in it is literally zero," says Jarrod DelPrado (you'll meet him and his brother Justin later on in this chapter). "We're just two guys building, and for the past several years we've had many bikes on order."

That should tell you something: listen to your own voice and you just might create something as sought after as they have.

This question is so complicated I had to read it several times before I knew what you meant. I would simplify to:

But what's to say of the trends in motorcycle design? The ones that probably brought you to pick up this book and thumb through it in pursuit of the golden insights of café racer or sport bike design? Well, that's a tricky question. To answer it, let's take a look at how we got to where we are now.

HOW WE GOT HERE

As in virtually all things, there have always been and always will be trends in motorcycle design. This book is evidence of that truth. To understand why this is not an instruction manual for the design of one particular style, consider the recent past.

The twenty-first century has, so far, been a strange time for motorcycle designers. When we kicked off the new millennium, choppers were getting an unusual amount of love for a style more commonly associated with havoc than with corporate-branded theme bikes like those being churned out by the guys at Orange County Choppers, whose internal family drama was strewn all over reality TV. Then a few things happened that seemed to cause $50,000 choppers to lose their sex appeal.

First, the United States economy took a dive. If you wanted to throw your leg over something stylish, it suddenly became much less likely it would be an expensive, built-from-the-ground-up custom chopper. American culture needed something cheaper to latch onto that could still provide that sense of individualism. We turned our eyes to the past and found a sea of vintage bikes just lying around, waiting to be reborn. For just a few hundred bucks, anyone who wanted to could pick up a Craigslist junker, overhaul the engine, strip off a bunch of parts, flip the bars upside down, and hit the streets. Motorcycling suddenly became a whole lot more accessible.

Second, technology happened.

In a very short time, technology made some significant leaps. Personal computers became fast and affordable laptops. CD players became MP3 players. MP3 players became phones.

Phones became pocket-sized computers more capable than any desktop computer made just five years earlier. Then laptops became tablets. While this may seem irrelevant to motorcycling, there is an aspect of all this technology many of us just haven't been able to resign ourselves just yet: the black-box-ness of it all.

For all the smooth lines and invisible hardware and glossy perfection of our newfound personal tech, there's some small part of our brains that doesn't trust it. We can't open up smartphones or tablets to examine their guts. We can't stand in awe as we watch our children take them apart and put them back together the way we once did with radios and remote control cars. We can't even replace their batteries. We carry these items around with us and use them every hour of every day, but we feel cognitively disconnected from them.

"As we become more and more modernized," says Alan Stulberg of Revival Cycles (you'll meet him soon too), "everything becomes more of a black box. You don't know what's in it. The iPhone is the perfect combination of aesthetics and function, but there's not a lot of soul in a piece like that. There's just not a lot there. No one knows the story behind how it's built because they purposely keep it secret."

It's just too hard to bond ourselves to something we can't peer into. It's our hardwired urge to want to be able to understand the things we use. We want to see into the soul of the thing. Minus that ability, technology becomes foreign. Alien.

"In history," Alan continues, "I think products were produced in an honest, out-front way because they had to be. Meaning that when someone said, 'Put shoes on my horse,' you watched them put shoes on your horse. Now, you don't know how the tires on your car are installed. You don't know how anything is made. Everything on your car is covered with plastic. You have no idea. A return to vintage style, the revival of artisanal, handmade, hand-built ways, and the presentation of that, is attractive to us as humans because we want to know where the shit we have was made—where it's made and how it's made and what makes it what it is."

Motorcycles are an answer to that. Compared to the magic and enigma of a smartphone, motorcycles—and vintage bikes, especially—are boneheadedly simple. They're primitive. They have souls. They have stories.

"I think people will always be obsessed with how the thing came to be, because that's the soul of the thing," says Alan.

Individualism, identity, connection, soul. Aren't these the reasons why you picked up this book? Motorcycles are so much more than motors on two wheels.

Our love for them is about much more than trends.

THE STYLES

The builders featured in this book are poster children for individualism. There isn't a typical designer among them, and they don't tend to build bikes you could easily group into one style or another. To create a frame of reference, however, and to give you a solid view of the landscape of possibilities, it's useful to take a look at the different bike styles that have emerged over the years. When you begin your own build, you may find hints of some of these styles bleeding into your own work, and you may not. It doesn't matter. The following descriptions are included just to give you a sense of the box from which you'll try to free yourself.

NAKED

Naked bikes became "naked" when sport bikes became popular and the standards suddenly looked stripped down and minimal by comparison. Of course, this BSA is anything but standard. *Lee Klancher*

Naked bikes are an evolved version of the standard motorcycle. Standards were the no-frills version of a motorcycle model, with upright seating positions and minimal or no fairing.

Today, machines with a similar stripped-down style are known as "naked" bikes. Naked bikes feature an upright riding stance, with the rider's shoulders and hips in a natural position and the feet below the shoulders by way of mid-controls (foot pegs and pedals located near the horizontal center of the bike). If there's a windscreen, it's small. If there's a fairing, it's because someone added it. Naked bikes are usually cheaper than sport bikes and large touring machines.

Naked bikes are the flexible daily riders built for life around town, afternoon rides, or whatever you want to do on a paved road. Throw on a suit and it'll get you to the theater in style. Throw on a backpack and it'll go grocery shopping. Throw on an anachronistic mustache and your 1970s Honda will fit right in at the dive bar in the cool part of town.

CRUISER

Triumph's version of a cruiser features many of the same characteristics as one made by Harley-Davidson or anyone else: a relaxed position for a long day's ride. *Lee Klancher*

A cruiser is what happens when the owner of a naked bike decides it's time for more leg room. The foot controls are up front. The handlebar puts the arms more straight out. An increase in low-end torque makes for less shifting at lower speeds and easier handling.

These bikes make day trips a bit easier, but they can still bring on fatigue; sitting in an upright position or a slight recline for long stretches with the wind powering into your chest can be tough on the hands and shoulders. You also don't want to do any high-speed cornering; the low clearance of a cruiser and the deep lean required to take a corner at 35 miles per hour will leave a nice scuff on your exhaust or frame. These hindrances aside, cruisers are usually downright comfy.

If you want more than the typical 50 to 70 horsepower that comes with a cruiser, you can seek out a "power cruiser." Same style, but with a beefier engine.

SPORT

Leaning through a curve at Firebird Raceway in Arizona, Jarrod DelPrado shows off the qualities of a sport bike: aerodynamic plastics on a high-performance machine. *DP Customs, LLC*

They're uncomfortable. They get lousy gas mileage. They're ridden by twenty-three-year-olds with color-coordinated helmets and unhappy girlfriends. But if it's ungodly speed, acceleration, braking, and cornering you're after, a sport bike will get you there.

These guys are all about performance. They're often powered by inline-four or V-twin engines, and fairings are used from end to end to reduce wind resistance, even entirely concealing the engine. All that plastic results in a sleek, futuristic look that offers plenty of room for multicolored racing stripes.

The rider leans far forward over the tank and into the wind by way of a tall seat height, rear-sitting foot pegs, and low bar controls. At high speeds, this position is an asset. At low speeds, it can be a back-wrenching, wrist-straining affair. Despite common sightings around town, they're more ideal for a track than for hauling groceries. Like taking out a tuned-up NASCAR racer for a Sunday stroll.

If you're more the wolf-in-sheep's-clothing type, you can always hunt down a naked sport. It'll give you the stance and comfort of a naked bike along with the power boost you crave out on the open highway.

BOBBER

This bobber also belongs to Jarrod DelPrado, but it shows a very different side of motorcycling. Here, the goal is a bike that's as stripped down as possible so that only the attitude is left.
DP Customs, LLC

The bobber, if you're one to believe stereotypes, is probably the bike most likely to be seen fleeing the scene of a crime. Next to the chopper, no bike carries more built-in attitude or grit than the bobber.

The bobber was the first custom motorcycle variation to focus on getting stripped down. The key to bobber design is in removing any part of the motorcycle not directly involved with its performance. Purportedly this was originally done to reduce cost (and to reduce weight, which can improve performance). Today, bobbers are often built for the sake of style. They feature a minimalist appearance, which, according to purists, must include a "bobbed" rear fender. They are typically built on stock, unmodified frames, as opposed to choppers, whose frames are heavily customized or built from scratch.

Bobbers fall into the cruiser category due to their upright or laid-back riding positions, but they can also feature mid-controls, making them more like a naked bike.

ADVENTURE TOURING

Adventure touring bikes are like beefy dual-purpose bikes with luggage. Hit the highway or the dirt—it'll handle either one just fine and leave you plenty of storage space for your gear.
Lee Klancher

If you were filming a documentary in the Arabian Desert and could travel only by motorcycle, an adventure bike is the one you'd want to take. They have large-displacement engines, windscreens for weather protection over long distances, fairings to reduce wind resistance, big tanks to handle long hauls between fueling stations, an upright seating position, and a surprising number of ways to attach luggage.

Adventure bikes also are equipped with enough suspension travel to handle rugged terrain and street-legal tires with some knobs to give you some bite on nasty terrain.

Touring bikes are similar machines designed to gobble hundreds of miles of pavement per day. They have the same large fuel capacity and luggage, with a more streamlined look, shorter suspension travel, and more creature comforts.

Unless you're fond of the super-utilitarian look, adventure and touring bikes almost always land on the ugly side. They're tall. They have drab paint jobs. It's really rare that a custom motorcycle builder goes to work on an adventure motorcycle.

That's not to say it can't be done.

BAGGER (DRESSER)

Baggers are also great for hauling that extra gear, but they're strictly for paved roads.
Lee Klancher

A bagger is a prettier type of touring bike—the type you see on city streets, out on the highway, at pretty much any "bike night" in Arizona.

The bagger's name comes from its use of side bags—large saddlebags or hard cases typically made either of leather or plastic—so you can get around the limitations of motorcycle travel and stretch out for a longer trip. Baggers feature the upright riding position of a naked bike or cruiser, and can be found with either mid- or forward controls. More often than not, they have large front fairings, windscreens, large fuel tanks, enough torque to hit 80 miles per hour at under 3,000 rpms, a passenger seat with a backrest, and even a stereo.

It's the bike you take out for a long Sunday ride to a small-town roadhouse two hundred miles away. They're most likely to be seen in pairs, in groups on a charity run, or parked in rows outside the breakfast joints where charity runs start and end.

CAFÉ RACER

"Café racer" was the nickname given to bikes used to run impromptu races starting from English cafés (truck stops). Riders would throw down bets and whoever made the agreed-upon round trip before the song on the jukebox ended won the wager. Any modification to the bike was for the sole purpose of improved performance.

Modern café racers have the characteristics of a sport bike, but with vintage looks (they're usually built from vintage bikes). Take a naked bike, soup up the engine, flip the handlebar, strip off some parts, and you have the basics. Café racers also often have "rear sets"—foot pegs and controls that sit farther back on the bike so the rider can lean far forward.

Best ridden with your knee to the ground around a curve on an English highway.

The key to a café racer is to strip away any part that doesn't directly relate to performance, and then hide what's left, usually by way of a seat cowl, which conceals the battery and much of a bike's electronics. Add a pair of rear sets (rear-positioned foot pegs and controls) and you're off to the café races.
David K. Browning / E3

OFF-ROAD

Off-road bikes are all adventure and risk and dirt. They're built light and nimble, and meant to be ridden like a madman.
Lee Klancher

Off-road bikes come in a breadth of forms, including motocross, enduro, trail, trial, flat-track, track, and more. They're frequently smaller bikes with a high center of gravity, tall seat height, a strong suspension for handling rough ground, and knobby tires for maximum grip. They're built to climb over logs, cut through brush, and fly over hills without breaking a sweat.

If you need to forego the trailer and ride back home to the city afterward, there's even a dual-sport: an off-road bike with street-legal components—turn signals (if required by state laws), license plate, and headlight.

STREET TRACKER

Speaking of dual-sport.

In the mid- to late twentieth century, riders began putting lightweight dirt bikes to the test on oval tracks in a sport known as flat-track racing. It wasn't more than a minute later that flat trackers wanted to ride those same bikes home on city streets. Throw on a headlight, a license plate, a turn signal or two, and you've got yourself a "street tracker."

It's just as common today for one to be built purely for aesthetic reasons. In a clear nod to the racers who inspired them, some owners even top off the look by taping up the headlight, a trick used to keep broken headlights from spilling onto the track.

Much like off-road bikes, street trackers are built to slide sideways around turns, carve lines, and throw up dirt. They just happen to also have the street-legal requirements covered.

A Triumph Bonneville is a little on the hefty side for the kind of flat-track bikes this one is meant to pay homage to, but it does a great job of capturing the minimalism and stance typical to the style. *David K. Browning / E3*

NOT PRESENT AT TIME OF PHOTO

A minor disclaimer:

You can crossbreed almost any two of these styles into a hybrid. It would take a set of encyclopedias to cover the aesthetic components of every variant. While most of our featured bikes make some nods to the styles described here (as well as their kin and offspring), and although many of the builder-provided insights are relevant regardless of genre, there are a few branches of motorcycle design this book doesn't even come close to touching. This isn't due to classism or exclusivity, but because the builders who have shared their knowledge on these pages simply don't work in those arenas.

Namely, nothing in this book could be reasonably considered a chopper. There are no scooters, mopeds, or underbones. There are no purebred dirt bikes (though a few look like older and smarter siblings). There are no baggers. And if you're interested in touring bikes, you'll have better luck at a dealership.

If you're interested in awesome, however, carry on.

If you're interested in awesome, however, carry on.

BUCK IT

Between this array of styles and the fact that none of our featured builders strictly adhere to them, you should by now notice a theme emerging:

There are no rules.

Motorcycling is an exercise in individualism. It's about looking down at the asphalt beneath your feet, seeing those dashed lines blur past, and knowing you're living your life the way you want, on the machine you want to live it on. No style has hard and fast rules. They all evolve. They all have their pros and cons. They all work better in some environments than others. They all earn funny looks from bystanders in the wrong context. Pay no attention.

It's your ride.

Do what you want.

OUTLIERS OVER SUPPLIERS

It's this mind-set that makes it so difficult to walk into a dealership and find the exact bike you want. Even if manufacturers could keep up with trends, owners would still want to customize every detail. Individualism trumps trend, and individualism can't be churned out in bulk. To get what you want, you have to pick the parts yourself. You have to turn the wrenches. You have to twist the screwdrivers. In doing this, you achieve something no corporation can. If you're still on the fence about building your own bike, this is what you need to understand:

Only an individual can build something individual.

Only an individual can build something individual.

If a mass-ufacturer was to rise up to meet the current trend, it would have to be via a choose-your-own-adventure platform with a never-ending line of aftermarket parts and a YouTube series. No one is succeeding at this.

When it comes to flexibility, individuality, and even collaboration, artisanal motorcycle builders are in a better position to scratch a customer's itch than any manufacturer. Likewise, motorcycle buyers who ache for custom are going to have better luck with an artisan than the guy behind the parts counter at the Harley-Davidson dealership. He may know a thing or two, but he's not going to teach you how to weld, and he's not going to craft the one-of-a-kind red leather seat you'll be planting your ass on for the next two years.

It's simple: outliers over suppliers.

And that's why this book doesn't focus on a single style or trend, but rather a collection of unique voices in custom motorcycle design.

Yes, you will notice themes in the images in this book. Don't bother worrying about them. When a lot of individual choices are made, themes invariably appear. Trends are just an effect of those themes. They're a result. They only become homogenized when enough people stick to them.

Don't be one of them. Trend isn't the point. Individualism is.

This book presents the concepts, the ideas, and the elements of motorcycle design so you can find your own voice. It's not about picking apart what makes a café racer a café racer—it's about picking the brains of some of today's most iconic motorcycle builders to pull out a model you can use in your own garage to create the machine so distinctly yours that only you could have created it.

YOUR INSTRUCTORS

The builders featured in this book are poster children for individualism. There isn't a typical designer among them, and they don't tend to build bikes you could easily group into one style or another. To create a frame of reference, however, and to give you a solid view of the landscape of possibilities, it's useful to take a look at the different bike styles that have emerged over the years. When you begin your own build, you may find hints of some of these styles bleeding into your own work, and you may not. It doesn't matter. The following descriptions are included just to give you a sense of the box from which you'll try to free yourself.

Max Hazan
Geoffrey McCarthy

JUSTIN AND JARROD DELPRADO

DP CUSTOMS

Jarrod and Justin DelPrado (left and right, respectively), the brothers and best friends behind DP Customs, in their shop in New River, Arizona. *DP Customs, LLC*

Justin and Jarrod DelPrado are, in Jarrod's words, "Two brothers and best friends who, for the entirety of our lives, have loved anything with an engine." They were raised, however, to go to college and get nice, secure jobs with health insurance. Which they did. But then:

"Twenty years of our lives went by working in a corporate environment that makes it really easy to stay," says Jarrod. "Next thing you know, you're institutionalized."

When elder brother Justin customized an old Ironhead Harley and then sold it off, the bike's new owner commented on how nice the work turned out. When Justin moved on to other bikes and started making side cash, it didn't take long for Jarrod to feel the urge to join him and for the two to go all in.

Midlife crisis?

"It's definitely some sort of a midlife awakening," Jarrod says. "Why would you give what is supposed to be your passion your D-grade list of time when you're tired and burned out and you need to get your dry-cleaning ready for Monday?"

Both quit their jobs and started building motorcycles full time.

Before all that happened, though, they learned something else by way of their mother: how to be a clean freak. "She used to make us press our shirts to go bag groceries," says Jarrod. The lesson stuck.

These days, every last machine to roll out of their New River, Arizona, shop is as clean and pressed as can be. The brothers cite their self-discipline to stick to what they like. They hide cables and wires away while still leaving them accessible for maintenance. They avoid layering on design elements at all costs. "We see bikes on TV shows and think the motorcycle looks great when it's sixty percent done, and then when they finish it, we think it looks terrible," Jarrod states.

Though they've since branched out to other platforms, the DelPrado brothers made their mark with Harley-Davidsons, pushing the bounds of that V-twin base into myriad directions few would've thought possible.

"Justin liked bobbers, and I liked café racers," Jarrod explains. "We believed our contrasting styles and ideas could make some cool motorcycles."

Judging by the number of times they've been featured in the moto-world's most popular blogs and magazines, they were right.

If Jarrod and Justin have just one thing to teach you, it might be summarized as this: "Quit your job."

(Or perhaps "There are no limitations to what you can do with a good bike.")

If Jarrod and Justin have just one thing to teach you, it might be summarized as this: "Quit your job."

This lovely beast called Two-Lane is named for one of the DelPrado brothers' favorite movies, *Two-Lane Blacktop*. Note the squeaky-clean design sensibility here. There are no exposed wires and every detail is as smooth and polished as it can get.
DP Customs, LLC

ALAN STULBERG

REVIVAL CYCLES

A jarring tintype shot of Alan Stulberg, designer extraordinaire and cofounder (along with Stefan Hertel) of Revival Cycles in Austin, Texas. *Alan Stulberg*

The 1997 Ducati 900SS SP-J63 by Revival Cycles demonstrates the team's taste for fast looks, a visual connection between the tank and tail, and smart use of color (all things discussed later on in this book).
Alan Stulberg

For Alan Stulberg it started early. "I always drew things, and I liked riding motorcycles," he recalls. "I lived for that. As a child, everywhere we rode in the car—and my dad loved to drive because he had to, because he was in the oil industry—I would go with him on these road trips. And the entire eight or nine hours in the car, all I did was envision myself on my dirt bike, jumping every single driveway and every single field and every single hill."

Alan didn't realize until much later that he could turn his dual obsessions of motorcycling and drawing into a profession. Prior to making that connection, he was on a very different path. "I was doing investment planning for people," he says. "I did software sales. I went to school and got a degree in business as an adult because that was what I was gonna do: I was gonna have a serious career as a businessman."

Alan got a job right out of school. Then, as with so many builders, something changed. "I didn't fit. And I realized I was never gonna fit."

That's when the name Revival Cycles started bouncing around inside Alan's head. "I put myself through school buying and selling and fixing motorcycles. Why couldn't I build bikes?"

He approached his friend, Stefan Hertel, an engineer who was, at the time, busy designing medical device implants. Stefan had never built a bike before, and wasn't convinced he could. Alan disagreed. He'd once paid a visit to the shop of revered builder Shinya Kimura and had come away with the conclusion that if perfectly normal people could do amazing work, then maybe self-doubt was a waste of brainpower. Soon enough, Stefan folded and the pair got to work.

Just build.

Since then, the name Revival Cycles has become synonymous with taste and a relentless devotion to performance. The pair have been featured on *Café Racer TV*, *Jay Leno's Garage*, and all over the Internet. Their shop is in Austin, Texas.

Alan's perspective? Don't waste your time idolizing anyone.

Just build.

JARED JOHNSON

HOLIDAY CUSTOMS

Former snowboarder, spinal-injury survivor, and custom motorcycle designer Jared Johnson, founder of Holiday Cycles in Portland, Oregon. *Pierre Robichaud*

Jared Johnson had no shortage of opportunities to work on motorcycles as a kid, starting with the 1968 Honda 50 he shared with his cousin and his brother. "I've always had sketchy old bikes," he says. "I've never had a modern bike. My family didn't have a ton of money, so none of the bikes were new. They always needed work."

Yet when he was fourteen, it was a Volkswagen Beetle that had Jared's attention. "I didn't get my license until I was seventeen because the car wasn't finished." When he finally did finish it, it was in better shape than one might expect from the average teenager. "It turns out I built a show car," he says.

His concern with the car's appearance carried over into his daily life. "We lived on a three-mile dirt road with potholes," he says. "I would wash the car at home and I would wash it again at the end of the road before school. I constantly obsessed about my vehicle."

Although he'd been riding motorcycles since he was a kid, Jared had a long way to go before coming around to building them as a profession. Out of high school, he left home to pursue a professional snowboarding career. The pursuit spanned fourteen years but never quite amounted to what Jared had hoped for. At thirty-four, he was still an amateur with little money. When he decided to move to Portland, that's when everything changed.

"It was my fifth day after moving to Portland. I dropped my trailer and all my belongings off at my new place and threw together my snowboard and some stuff and went up to Mount Hood here in Oregon. The fifth day I was up there, I broke my back. So I showed up, dropped all my stuff off, and came back in an ambulance."

At one point, he temporarily lost feeling in part of his right leg.

"It was a T-7 burst fracture, so they had to take out the whole vertebrae, or all the pieces of the T-7, and they put in a fake vertebrae—basically a cage made from some of the bone marrow left over from that portion of my spine. Then they fused my spine with two eight-inch rods and eight standard lag bolts, which went into the remaining part of my spine."

While he recovered from the injury, Jared spent a lot of time staring at an old bicycle sitting in his apartment. As soon as he was able to move around again, he put a motor in it. This set the proverbial wheels in motion.

> "I started buying up crappy bikes for $300 or $400. I sold off a couple, and got re-stoked about motorcycles."

Breaking his back forced Jared to consider a hobby that would benefit him the rest of his life. "It was somewhat of a blessing in disguise," he admits. "It was a very big turning point in my life."

His friends nudged him toward customizing motorcycles. His dad, who always had an affinity for motorcycles and old cars, was also supportive.

"I started buying up crappy bikes for three hundred or four hundred dollars. I sold off a couple, and got re-stoked about motorcycles."

Swooping lines, matched curves, patina, and minimalism. Johnson brings these ideas to a level of high art. *Jared Johnson*

When it came time to build his own bike, Jared found even more inspiration from his youth. "On our property there had always been some, like, '50s Buicks melting into the ground, and I just always thought they were cool." The aesthetic style Jared is known for bends heavily toward the curvy, bold looks from the glory days of car design. He sees that roundness in the parts of a motorcycle—fenders, wheels, engine parts. "Instead of a lot of straight lines," Jared says, "I felt like complementing that. It's a timeless design."

It started with a rusty tank. Jared had done some work on a frame but wasn't satisfied. When he found the tank, he decided to cut up his prior work and build everything around it. "I suddenly felt sure of everything I wanted to do."

Even then he questioned his design sensibilities. "I first used a straight header with ninety-degree angles on it, and it just looked awful. But when I rolled all the exhaust tubes to match the frame, I was checking myself and asking, 'Is this getting too weird? Is this a cartoon motorcycle?' But I knew I just had to keep going with it because it was my gut instinct."

The bike that came out the other side is remarkable, as are those he's built since. "It's always one step at a time. I typically will stand around and look at the bike for hours throughout the day, and just kinda dream up stuff. It's a process."

Jared has been churning out his creations for a few years now and has made a name for himself throughout the community. He attributes this partly to the Portland lifestyle. "Portland is a very DIY, very communal place," he says. "It's not really cutthroat. I've gotten calls from other shops who have told customers about me."

MAX HAZAN

HAZAN MOTORWORKS

Master motorcycle artist Max Hazan, the man behind Hazan Motorworks in Los Angeles, California, with his Ducati 900. *David Hans Cooke*

Max Hazan stands out even in a league of standouts. It's difficult to talk about his work without using a few superlatives. Even the esteemed builders in this book have expressed their admiration for his work. Any attempt to describe it is useless—the photographs will have to do that job. His story, though, is very human.

Eighty or ninety miles an hour on bicycle parts?

Much like Jared Johnson's, it starts with an injury.

"I'd just had a pretty bad crash on an enduro ride, and I couldn't walk for three months," Max recalls. "I was sitting on the couch and I had this vintage bicycle on a shelf in my living room. And I was just staring at it for like three months while I was recovering."

The bicycle was a beach cruiser. When he recovered, Max built his own frame and put a cheap engine in it. "It was made mostly out of bicycle parts. It went so fast that I actually got scared using the bicycle parts, because it was going eighty or ninety miles an hour. So I built one using motorcycle parts."

Eighty or ninety miles an hour on bicycle parts?

"It looked pretty cool, but it was terrifying."

Before the beach cruiser, Max had a

contracting company in Manhattan doing interior design and custom cabinetry, but there had been hints of his next career even before the enduro incident. Namely, he had put a dust curtain up to section off part of the workshop for metalworking. "It was a tiny spot with no windows." After the first bike sold, Max's dad asked him if he thought he could make a living building bikes. Max conceded that while it probably wouldn't make him rich, it might make him happy. "I never got into café racers or any of that vintage bike thing," he admits. "I was just a mechanical person. I just liked building motorcycles."

Eventually Max sold all the woodworking stuff and turned his tiny, windowless Manhattan space into a bike shop. Since then, it's been all magazine covers and superlatives. (Max eventually met a girl and moved his shop to L.A., but that is an entirely different story.)

For Max, it's not the individualism of a motorcycle that matters. It's the building. The process. "I was just always about building things," he says.

Besides showing off his exemplary design skills, this Royal Enfield shows off Hazan's woodworking know-how and earned him a truckload of positive press (granted, this happens to him a lot).
David K. Browning / E3

JOHN RYLAND

CLASSIFIED MOTO

John Ryland in the Classified Moto shop in Richmond, Virginia, sporting a branded T-shirt and propping himself up on a signature Classified bike.
Adam Ewing

John Ryland is on the other end of the spectrum from Max Hazan, but not because his bikes are any less distinctive. It's how he arrives at them that makes his story a remarkable contrast to Max's. It's also a story that should give you all kinds of confidence in your ability to put together something incredible in your garage.

"I'm constantly having to remind people that I don't know what I'm doing," he says. "Really, I just know what I like."

John doesn't sketch. He doesn't rebuild carbs. He doesn't fabricate fuel tanks. Not that he's incapable; he used to do all these things. He's broken stuff. He's chased down wiring problems. He's spent unanticipated cash when things have gone wrong. He just doesn't happen to enjoy these particular aspects of motorcycle building.

"I love the design process," he says, "and I love the ride, and I like taking this thing no one thought had any potential and doing something really cool with it."

To that end, what John does do is enlist the help of talented, capable people who will handle the things he either can't do or doesn't want to get into himself. Classified Moto is not John Ryland alone, but a team. When critics question his self-professed so-called shortcomings, he has this to say: "I don't have to know any more. I can just be honest about what I know and what I don't know, and be really proud of the product when we're done with it."

"I love the design process," he says, "and I love the ride, and I like taking this thing no one thought had any potential and doing something really cool with it."

It's an approach that frees him up to focus on the parts of the process he does enjoy. And it has earned the company a whole lot of fans.

John's signature style is hard to mistake. It commonly features gold forks, metal mesh for things like side covers, and the mono-shock

treatment, which gives every bike the shop touches a look that is all Classified.

This is very much intentional.

Prior to taking on his first build, John worked at an advertising agency for eleven years. When he got laid off and used his severance to start building bikes full time, he took a lesson or two with him.

"I saw that it was important to have a brand people can recognize. If you have aspirations of having your own business, it's just something you've got to think about."

It appears to be working. John and the Classified Moto team have so many orders at any given time that they often have to stop taking them.

Beyond bikes, Classified Moto does advance digital renderings for customers who want to mull over a custom design before pulling the trigger. And if you want something from the Classified shop but can't commit the cash for a whole bike, the team offers lamps made from junkyard finds.

If anyone can serve as a beacon of inspiration, it's John. And if you're worried about the haters, don't be.

"Go do it!" John says. "Don't worry about what some old guy said. You're not designing it for him. He's not gonna ride it. You're gonna ride it. Do what makes you happy."

The XV920R6, which was Classified Moto's first customer build, features several signature moves, including a front-end swap and a mono-shock treatment. They've pushed this aesthetic in every direction since then, but this bike started it all.
Adam Ewing

FEATURED PHOTOGRAPHER:
DAVID K. BROWNING
E3 SUPPLY CO. VINTAGE MOTORCYCLES AND ACCESSORIES

If it's great motorcycle photography you need, David Browning is a man you should know. His shots appear throughout this book, and he has been an immense support in the process of creating it.

His story is a wild one.

Prior to starting E3 in Brooklyn, New York, Browning worked in music and fashion photography. In 2010, he built out his first motorcycle, a 1976 CB550. He did this in a fourth-floor East Village apartment (so if all you have is a tiny garage, well, you have a lot). Soon that build was featured in a post on the wildly popular Bike EXIF blog, and David was commissioned by actor James Ransone (*The Wire*, *Generation Kill*, *Treme*) to build an exact copy.

For his next feat, Browning joined up with the high-design motorcycle lifestyle and culture magazine *Iron & Air* as a photographer and content contributor, where he was responsible for two cover shots and seventy-eight pages of published work in the magazine's first year alone, documenting the New York custom moto scene in the process.

From there, he moved into a bigger shop. Then an even bigger one. Then he formed E3 Supply Co., adding handmade leather goods, moto accessories, and custom watches to the roster. He currently splits his time between bike builds, photography, creative direction, and moto- and photo-related leather goods. Browning makes everything that leaves the shop on his own, whether finished bikes or accessories.

Somewhere along the way he met Max Hazan, one of this book's featured builders. He even moved into Hazan's Brooklyn shop for a while after Hazan relocated to Los Angeles in 2013.

Browning's clients, whether for bike builds or creative services, include Bike EXIF, Metzler Tires, Triumph America, Dianese, Virgin Mobile, Jawbone Electronics, and B&H Photo. His projects include a partnership with Google to create handmade leather watchbands for Android Wear devices.

You can learn more about Browning at www.e3motorcycles.com and www.instagram.com/e3motorcycles.

Featured photographer David Browning hangs out next to some two-wheeled perfection of his own creation in Brooklyn, New York. His shots can be found throughout this book. *Julien Roubinet Photography*

ZEN AND THE ART OF DESIGN

With such diverse perspectives, you'd expect these builders to have very different beliefs about design. But while each has distinct opinions, all are very much built atop a few commonalities.

First, although you might expect to walk over to the desk at a custom shop and find a stack of sketches, equations, and colored pencils, for our builders this is not the case. Rather, they say, motorcycle design tends to happen organically. It can develop around a single part, and you never know what part that might be. "It can start with anything," says Max Hazan. For him, it was once a headlight. On another occasion, it was the rear tire from an old Indy car. But, he says, "It usually starts off with an engine. I think, 'I'd love to build a bike around that engine.' I put the engine on the table and I kinda look at the engine and see what wants to go around it. What shape the bike should take."

By all appearances, motorcycle design is more cooking than architecture. Just as a capable chef can build a delicious, unique meal around a single ingredient that happens to be in stock that morning, a motorcycle builder can, and often does, rely solely on his wits to get him to the end. Rather than pages of planned, perfect lines, most of the builders featured on these pages navigate their way through one choice at a time, are often surprised by what happens along the way, and are thrilled with how it comes out. It's not usually clearheaded vision that guides the process. It's more seat-of-the-pants decision making. Like riding itself, it's present. It's immediate. It's now.

When it comes to up-front planning, Stulberg is the standout. "I'm thinking about the end before I start," he says. "I have lots of sketches on napkins. But then many details change mid-stream because this makes more sense than that."

EMBRACE THE CONSTRAINTS

Second, there's no such thing as a blank canvas. "You start with a set of constraints," says Stulberg. "All we do is solve problems based on an aesthetic preference."

It's important not to see this as a downside, however. Design in any medium requires constraints. It's the things you can't do

The source of inspiration behind some of Jared Johnson's most notable builds was this Schwinn bicycle, which now hangs on a wall in his shop in Portland, Oregon (and sometimes doubles as a hook for his headphones).
Robert Hoekman Jr

that dictate what's possible. For example, a motorcycle wouldn't be a motorcycle if it had four wheels and wings. By putting a motor on two wheels to create something you can safely call a motorcycle, you've ruled out billions of possibilities. It's a good thing.

Then there's the challenge of dealing with exposed parts, something you don't have to consider so much with a car or a house. "Motorcycles are great—especially vintage," Stulberg says, "because they have to wear their design on the outside. The tires, the chain, the suspension—it all has to be out in the open, and if you want the bike to be pleasing aesthetically, all of that has to be taken into account. I don't think they're that difficult, though, because you're still dealing with the fact that you have two wheels, it needs to fit a human body of general size, and it has to have a certain function."

In other words, it's not like staring at a blank piece of paper in a typewriter for hours on end, panicking over what words to write. With constraints, you have a place to start.

The DelPrado brothers begin their customer conversations by narrowing down these constraints. They start with the "ergonomic riding requirement" of the person who will ride it—the person's height, weight, as well as riding style and frequency. For their Rusi bike, based on a modern Triumph Bonneville, the customer requested they build whatever they most wanted to build. Even then, Jarrod tried to find the edges: "He said he was going to ride the hell out of it on the streets of New York. So right there, we knew we wouldn't be doing a hardtail."

AIM FOR CLEAN

Third, clean design is good design. Spend a little time reviewing any design considered high quality, and you'll almost invariably find it is composed of only the most necessary elements, arranged and presented in their best form. Sticking to this principle will help keep you out of danger. It'll keep you from tacking on that extra gadget that doesn't need to be

following page:
Max Hazan's Supercharged Ironhead
Geoffrey McCarthy

there, from using one paint color too many, from landing on the side of garish.

"Try to say as much as you can in as few words as possible," says Max Hazan, channeling Ernest Hemingway. "Use one line when one line is all you need, but make that one line as pretty as you can."

John Ryland echoes that perspective: "It's sorta hard to argue with a motorcycle that looks like everything is serving a purpose, but is all neat and tidy and looks athletic and really capable."

No matter how complex the bike at the beginning of the process, stripping away the layers to reveal the simplest form of the machine invariably makes it a more beautiful thing at the end.

When in doubt, minimize. Be ruthless about it. Whatever is left, make it look as good as you can.

The act does come with its challenges, however. Simple can be hard to build. As Hazan points out, "I make things hard for myself because I think of these designs that are super minimal and really clean, and those are usually the hardest to make."

Try not to see it as a blocker. Overcoming the challenges will make the end result all that much more worthwhile.

FORM FOLLOWS FUNCTION (SOMETIMES)

Finally, the way a bike looks is intrinsically tied to the function each part provides. And the way these parts function together, to some extent, dictates how the bike will look. Their shape, their fit, their position, their style—all of these things contribute to the motorcycle's appearance.

Max Hazan sees this throughout history: "On all the really classic bikes, all the bikes that have been iconic, it's been one hundred percent form follows function. I've found that when form follows function, it usually looks aesthetically pleasing. A part that's shaped to function looks good inherently."

When the parts come together to form the whole, he says, "You could have a really busy design, but as long as the pieces function, it doesn't look cluttered."

The axiom is not, however, a universal truth. It depends on what you're trying to achieve. For some of our builders, for instance, beauty takes priority. Jared Johnson explains, "There's a lure to bikes being messy and dangerous and not so polished and smooth. I would rather a bike look really sexy than have all the proper parts. I still always put front brakes on my bikes. But I would rather build a rolling piece of art than something that's more about function."

It can also depend how you're going about the process. Hazan's designs tend to come from functional shapes. He doesn't dream up a bike "that looks like a rose petal"—he conjures up functional parts that fit into some aesthetic idea and then works backward to figure out how to make them. "Form does follow function," he says, "but a lot of the time function has to follow form for me too."

However you get there, your voice will come out. A maker will almost always inject a unique style into the things he or she makes because people, in small and endless ways, are inherently unique. We have distinctions. We have nuances. Yours will come out no matter what. Copying elements is by all means encouraged—every creator finds new ideas through copying others—but if you forego the urge to design every single detail just as someone else already has, your individualism will come out. You'll see yourself in the result. Others will see you there as well.

You can't argue with that.

Firestone

219

THE BONES

There are places where motorcycles go to die. Some slink to their last sputtering breath in that two-wheeled cemetery known as the motorcycle junkyard. Others gasp their way to the back of a garage, left to rot behind rakes and push brooms and dusty boxes of heirlooms. Others aren't dead at all, just pushed aside by fearful spouses or displaced by newer, shinier versions.

previous spread:
David K. Browning / E3

following page:
Geoffrey McCarthy

No matter their state, owners far and wide end their relationships with them with a ritual. They wipe that chunk of grease out from the corner of the transmission. They make a mental list of their most recent attempts at maintenance. They write down the modifications they've made over the years. They dig the title out from the box in the closet. They run a hose and rag over the frame, the seat, the wheels. They pry the keyring apart and yank the key from its death grip for the last time. They click the "Post" button on the Craigslist ad, their breakup with the beloved machine now a public announcement, their years-long love affair reduced to a written description and a price tag appended with the words "Or Best Offer."

It's a sad thing, this ritual. Even the most ruthless among us will feel a tinge of future regret just thinking about giving up our lovely beasts. But it has to be this way. For progress to be made, for new life to be found, it has to be this way. Every owner has to say good-bye at least once or twice.

Good.

Because your custom project has to start somewhere, and someone else's cast-off crush is the best place to start. They're cheap, they have histories, they have memories, and they're dying to be brought back to life.

They just need a chance. And maybe a new carburetor.

This chapter is about where to find the one you'll call your own and what to do with it once you get it into the garage.

DUCATI

THE DONOR BIKE

A 1200cc Harley Sportster donor bike waits to be stripped down at the DP Customs shop in Arizona. This bike will undergo a major transformation, eventually becoming The Player.
DP Customs, LLC

It starts with a donor bike.

Doesn't it? It's a nice term, "donor bike"—all full of promise and low on price. It's what builders call the bike they start out with, the one that will become something else after a few weeks (or months) under the shop lights. But despite the name, donor bikes aren't often free and their innumerable possibilities and conditions and styles and shapes all have an impact on what rolls out after the thing finally has life breathed back into it. So, really:

It starts with a search.

"Common bikes usually come from Craigslist," says Max Hazan. "It's local, so you can see the bike in person, as opposed to most eBay auctions."

John Ryland seconds the motion, but ups the effort: "More often than not, we find our donor bikes on Craigslist," he says, describing Classified Moto's typical procedure. "I have about a three-hundred-mile radius that I search and will drive to find bikes. For hard-to-find bikes or bikes I need on a tight timeline, I'll drive farther or I'll have the bike shipped to me. I hate to buy things sight unseen, but sometimes that's the only way."

Jared Johnson is an even more extreme Craigslist user. "If I'm on a road trip, I'll stop in random towns and check their local Craigslist," he says. "I've come back to Portland before with a truckload of bikes from San Diego, San Francisco, Northern California."

Jarrod DelPrado, likewise, says that while he and his brother find most of their donors in their home state of Arizona, they've pulled some from Nevada and California.

Revival Cycles is the exception here. Alan Stulberg points out, "Most often, our clients bring them to us. They come from all over the world, really. We rarely actively search for them."

The lengths you'll have to go to depend, at least in part, on how flexible you are about the particular bike you're willing to pull into the garage. Maybe you want to shine up a sibling of that long-lost love of a motorcycle you beat to hell and back during your last two years of high school. Maybe you really like the one the guy down the street has and you have a few ideas about how to make a better one. Whatever the case, the more specific your hopes and dreams, the more difficult it may be to make the first step. A broader range of options can mean getting started sooner.

Although Classified Moto now looks mostly for specific bikes like the Honda Nighthawk 750 and the XR650L—"fairly modern bikes with plenty of new parts available"—that wasn't always the case. "I used to look for pretty much the cheapest bike I could find," Ryland says, "and I didn't care if it was exceedingly ugly or uncool. For personal projects, my goal is often to turn an ugly duckling into something interesting. I didn't really care if it was shaft-drive or raked-out or looked more cruiser than café. You might consider using a single or twin for your first project. Fewer carbs can be a good thing when you're doing your own wrenching."

(On an aesthetic side note, Ryland says he also prefers big engines: "Motor-wise, I like the look of a big four-cylinder motor stuffed into a smallish frame. While a V-twin is often thought of as a muscle power plant, it looks surprisingly skinny from the front.")

Okay, so cheap is good. But then, it's also relative, and past a certain point, the quality starts to affect the process. Ryland continues: "We rebuild all our motors, so we'll still buy a bike that's blowing some smoke or down on compression a bit. But for the garage builder—especially for a first project—I highly recommend getting something with no major engine work required. It's a real drag to be able to realize your aesthetic vision only to be bummed when your friends complain that you're smoke-screening them."

INSPECTION TIME

So how cheap is okay? Will you know it when you see it? Our builders have some hard-earned opinions.

"We're mostly concerned with the treatment and mileage on the bike," says Jarrod DelPrado. "It's less about finding a great deal and more about finding a well-cared-for machine with low mileage."

Ryland adds some detail. "Check the charging system. Make sure the engine is sound. Make sure it shifts smoothly through the gears and doesn't pop out of gear."

Jared Johnson at Holiday Customs isn't so picky. "After buying random bikes through the years, I have figured out what bikes work better for the design that I have in mind. I buy bikes 400cc and up, single- or twin-cylinder. I'm not really into four-cylinder bikes. The engine does not have to be running or clean, just together. Rebuilding the engine takes care of all that. If the motor turns over and can roll, it's usually okay with me. If the price seems high, I will do a compression check, check for spark, shake the wheels to check the wheel bearings. If that stuff doesn't check out, I lowball their ass—ha-ha! I usually don't buy bikes over a thousand dollars, so I'm usually buying bikes that have issues I take care of."

That said, Jared does like to play favorites: "I've been known to buy and build XS650s and CB 450s. I really like the look of the engine. Honda and Yamaha were trying to compete with the other beautiful engines at the time, like Triumph."

A little Web research uncovers some more great advice: inspect the bike with your hands and eyes first, then your ears.

When you walk up to the bike for the first time, touch the crankcase with your bare hand. If it's warm, walk away. A whole host of electrical and other problems can hide themselves in a bike that's been running for a few minutes. A seller who warms up the engine before you arrive is a seller who shouldn't be trusted.

What's that he's telling you? He just ran it to the corner to fill the tank? How nice of him!

Walk away.

If the engine is cold, however, then dig in. Look for wear and damage to the sprocket, chain, forks, and any other moving parts you can put your hands on. Parts that need to be repaired or replaced make a donor bike cost a lot more than the asking price, even if you're doing all the work yourself. This includes worn tires, fuel tanks with rust on the inside (this is fixable in a lot of cases, but a hassle every time), damaged wheels, and any number of other things.

Are there scratches on the crankcase? The ends of the handlebar? The foot pegs? The mirrors? Anyone who's ever laid down a bike can tell you these are telltale signs that a bike has touched ground before. And a bike that's been downed can have problems you can't see from the outside.

If all is well, then you can ask for the key.

Beyond these signatures of good and bad,

An E3 Motorcycles build is used to point out an array of potential parts to look over when considering a donor bike purchase.
David K. Browning / E3

there are also some intangibles to consider. Namely, the person holding the key.

"The main thing that I do is evaluate the seller," Hazan says. "I take a look at how he keeps his house, his shop, and even himself. Nine out of ten times it's a pretty solid indicator of how the bike was kept. There is no way to open the engine up on the spot, so if the seller's place is a disaster, or they are a mess themselves, I usually walk away. I have bought dozens of bikes, and this has been pretty spot on. I have made a few bad purchases, though."

Jarrod DelPrado agrees. "When we find a bike that looks good in an ad, but then we pull up to a crappy house with an owner who clearly doesn't maintain anything he's got, we just keep driving. We've got plenty of experience buying bikes, and it's become easy for us to recognize detail-oriented folks who care for their machines properly."

And it's not just how the person or shop looks that matters.

DelPrado continues: "You can usually tell an honest seller from a grifter. If you feel good about what they're telling you and you want the bike, then go with it. Don't let a hundred bucks stand in the way of a bike that you know fits the bill. On the flip side, do yourself a favor and avoid bikes whose owners come across as assholes in their ads. As a general rule, if the ad has the phrase 'Don't waste my time' in it, the seller is an asshole. Avoid."

1 / Check for rust inside the fuel tank, as this can be difficult to fix or replace.

2 / Look for damage to the handlebar ends, mirrors, and foot pegs, indicating that the bike may have been laid down.

3 / Look for leaking fork seals, and to see if the forks are bent or rusted.

4 / Touch the crankcase immediately, as a hot engine could indicate that the owner had to warm the bike up before your arrival.

5 / Check the oil level, and for signs of leaks. Also, gunk visible through the viewing window can mean problems inside the engine.

6 / Check for a worn sprocket.

7 / Look for wear to the tire treads that could mean they require replacement.

BREAKING IT DOWN

You've gone through the Craigslist ads. You've visited a few bikes. (You didn't buy the first one you saw, did you?) You've thrown down the dollars, signed a title, and hauled the thing back to your driveway (then ridden it around the block a dozen times despite not having the new plate yet). It's Saturday morning, the kids are at Grandma's, and there's nothing more in the world you'd like to do than put a wrench in your hand and start hacking away. You're well past the part of the process wherein you fear the enormity of what you're about to do. It's time to get to work.

You face the machine. You take a deep breath. You think. You wonder. You wonder some more. Your chest sinks. Your brow furrows. It doesn't take long before you're staring down the barrel of that big question everyone who's ever built a bike before faces but you've forgotten to ask:

What do I do first?

STRIPPED FOR PARTS

There's no best answer to the question, just common preferences, and they run the gamut: start at the front, start at the back, start by running the engine, start by disassembling the engine, start here, start there. The only consensus seems to be:

Get to zero. Get it down to the bones. See what you're working with.

The art of a custom project isn't in its original design, but its future design, and its future design is hiding underneath all that crap the manufacturer put on the bike. The excessive plastic. The gaudy tailpiece. The air box. You have to chisel away at the stone and find the sculpture within. Strip the machine down to its birthday suit and see what you're really working with.

"Once we get the bike to the shop," says DelPrado, "we do a hundred-percent breakdown so we're left with the frame and the motor on a stand. We ditch pretty much everything else."

Jared Johnson points out that as tedious or daunting as it may be, this process can be a good time. "When I get a new bike, it's usually a pretty fun day," he says. "I put it up on the lift and start taking off most of the parts to see its frame and check out the lines."

If you want to try to predict the level of engine trouble you might have after you've got the thing back together, take John Ryland's approach.

"When we get a bike in the shop for a project, we first get it running pretty well—or attempt to—even if we know we'll be rebuilding it," John explains. "It's just good to know what you're dealing with. We don't spend a lot of time tuning it at first, but rather, we're just making sure there are no fatal flaws—a shot transmission, crack in the motor, et cetera."

One key piece of advice before you get too far: hit up the Clymer website and order the manual for your exact donor bike. If you're new to motorcycles, resist the urge to do too much damage until the UPS lady knocks on your door.

When you have the manual in hand, it's time to break stuff.

A bike sits in Max Hazan's Los Angeles workshop, waiting to be stripped down and turned into something amazing. *David K. Browning / E3*

It doesn't matter where you start as long as you can keep yourself organized, but you might find it helpful to take one section at a time. The seat can be a great starting point, because the second you remove it, it looks like you've done something important. Next, drain the fuel tank (you can find fairly universal instructions online), remove the tank, and set it in a place where it won't get scratched. If it's already beaten to death and you like it that way, then don't worry about it.

From there, you can remove the battery, which needs to be done before you mess with anything electrical anyway. Next, work your way through the mirrors, turn signals, gauges, headlight, handlebar, front fender, and anything else that's detachable from the front half of the bike without touching the forks (don't mess with those before watching a good long YouTube video on the subject).

As you do this, keep a few plastic bags and some masking tape handy. "We bag and label key fasteners that we know we'll be reusing," Stulberg says, "like engine bolts and whatnot."

Have you dirtied your hands on other builds in the past? Then you may be able to risk a few forgetful moments while staring at an unlabeled pile of bolts. "After rebuilding or disassembling this many motorcycles, we generally know what we're holding regardless of whether it's labeled or not," Stulberg points out.

But not everyone has done enough of these to identify a part from five feet away with one eye closed. Better safe than sorry. Grab a Sharpie, throw that collection of screws and fasteners from the handlebar gauges into a plastic bag, stick on a piece of masking tape, and write something on it you'll be able to decipher later. Remember, it might be months before you look at it again.

To fortify your future memory even more, pull out your iPhone and take some pictures. In three months, a photograph of which connectors went where could be the difference between a quick assembly and a week of self-education in motorcycle wiring.

following page:
The DelPrado brothers carefully tape and label a sea of wires for later reorganizing. By the time the brothers are done, not one of these wires will be left exposed. *DP Customs, LLC*

THUNDER MAX
AUTO-TUNE MODULE
THUNDER MAX
EFI MODULE W/ AUTOTUNE

A-11
GOLD

The tail section is another good place to go. You can do a lot there without getting yourself into trouble. With the seat off, you can more easily access the taillight, turn signals, rear fender, chain guard, and that enormous fairing you want to replace with a custom seat pan.

While working the back half of the bike, you may find it necessary to remove the exhaust pipes. If you do but you don't already have extensions for your ratchet set, now is a good time to head over to the hardware store. Exhaust headers can be tricky to remove and even trickier to put back on. Beyond that, about a hundred times throughout this process you'll be glad you bought good ratchet extenders.

Once you decide you need to remove the engine, you need to make sure you keep your thinking straight. Engines are full of oil. You need to remove it or else you're likely to end up with most of it soaking into your concrete floor. Besides that, you will have a hard enough time lifting the engine out of your bike, never mind one full of liquids. Before you try, drain the crankcase and transmission oil. See your manual for instructions.

Max Hazan gathers the goods to assemble one of his masterworks. This collection of parts is going to get a whole lot of media attention once it's organized.
David K. Browning / E3

PLANS AND ACTIONS

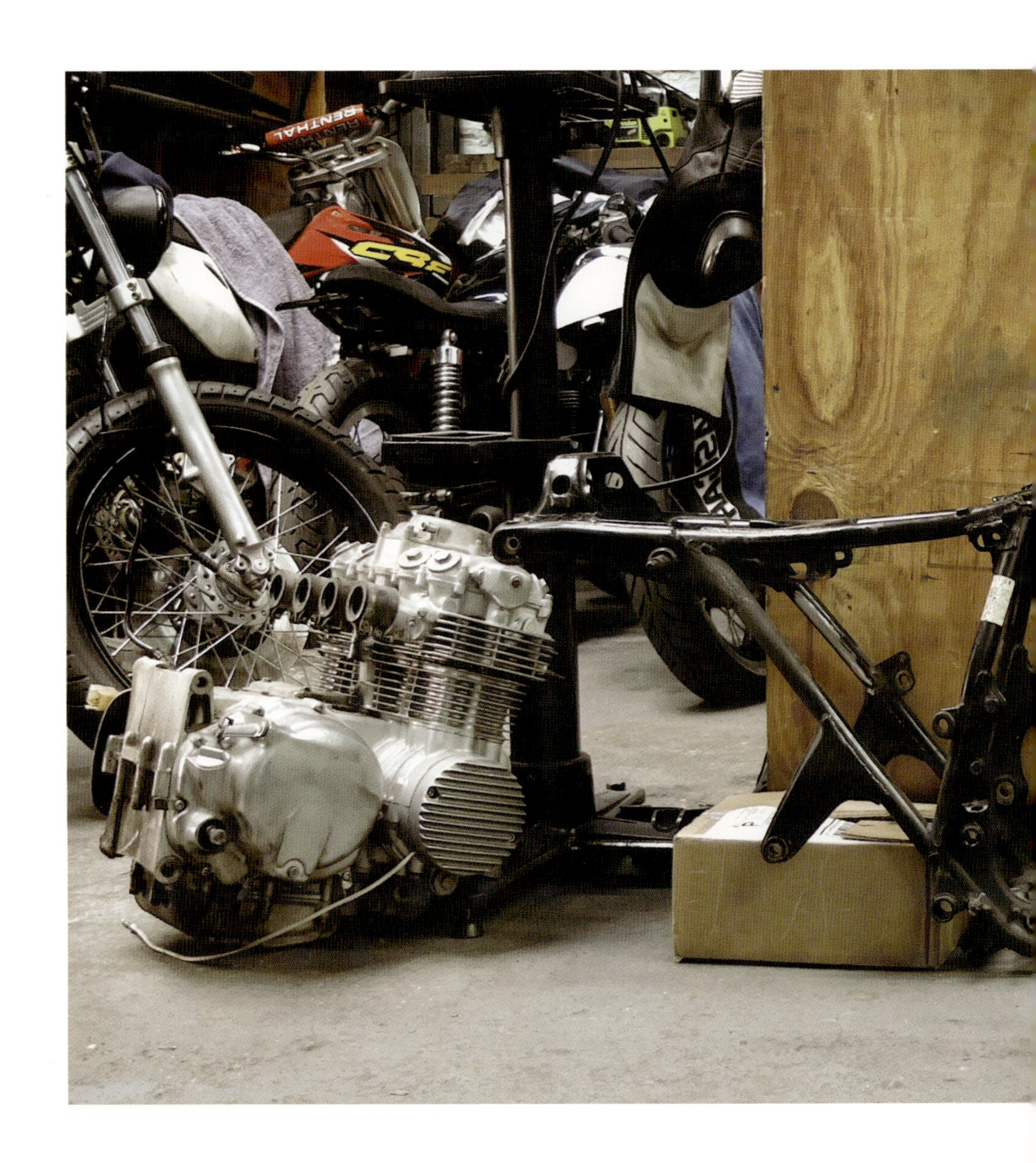

Once your naked bike is as naked as it can get, it's time to stand back and daydream. And scheme. Like everything else, this happens in a variety of ways and has several angles. First comes the staring.

A donor bike is stripped down to its frame. Now it's time to sit and stare until the ideas start to flow.
David K. Browning / E3

"I mostly stand around and look at a bike for hours once it arrives," Alan Stulberg says. "I'll try to see what perhaps the original designer had in mind and then assess what I might want to do differently. It's normally pretty easy to see what I might want to keep."

John Ryland and the crew at Classified Moto are also deliberate. "We try to plan frame modifications, tank work, and suspension and things like that in such a way that the bike is rolling when it needs to be," he says. "In other words, it's good to keep the big picture in mind. There's no use rushing to get a set of shocks when the swing arm is being powder-coated. There's also no use rushing to get your frame painted only to find you still need to weld three different brackets to it later. Take your time and don't be afraid to have the bike assembled and disassembled a bunch of times during the project."

"In other words, it's good to keep the big picture in mind."

DP Customs tends to shape a few ideas beforehand. "It probably sounds odd," says Jarrod, "but we've usually got ninety percent of the bike designed before we even get the donor. Most of the ergonomic and aesthetic elements are figured out during the design meeting with the customer. There will be minor changes along the way as the bike evolves, but we usually can see what it's going to become really early on."

Next comes the decision of what to do with all the parts now littered about the cutting-room floor. "If you're a motivated sort," says Ryland, "you can sell unused parts on eBay or wherever to offset some of the costs. Do a little research and see if the stuff you're junking is worth anything. If it's worth the time, sell it. Otherwise, do what I've always done: hoard that stuff and get rid of it when it gets really bad."

Many parts will find their way to the bottom of the trash can almost without a thought. It doesn't take long to decide you hate an ugly part. And after making the decision to toss the first one, it gets easier and easier. DP Customs doesn't waste any time at all anymore. Says Jarrod, "We start ordering new parts right away and hit up Advanced Metals for sheet metal, aluminum, and any tubing we may need."

Jared Johnson has a crafty way to save a few bucks by repurposing what might otherwise become wasted parts. "I trade all the parts I don't need with our local motorcycle parts guy for parts I might need," he says.

With a clean floor and a box or two full of labeled plastic baggies, you can focus on the design.

"I usually stare at it for a while," says Johnson, echoing Stulberg, "and start putting different tanks on it to get the creative juices flowing."

John Ryland elaborates: "Think of your project from all angles when it comes to aesthetics, and think of things you can do to play on your donor bike's strengths and mask its weaknesses. Tires are a big design element and can add a ton of character to the front and back of the bike as well as the profile. In general, the simpler the frame and more neutral the angles, the more conducive the bike is to customization, or in our case, 'Classification.' I love a nice straight seat line or one that kicks up slightly in the back. A low frame backbone is nice. Tanks with wide tunnels and simple lines are easier to work with. Your tastes will vary."

A welding job at Revival Cycles. The tunnel is a groove that runs lengthwise along the bottom of the fuel tank and determines how it sits on the frame.
Alan Stulberg

Does one aspect of a custom project matter more than another? Where do the guys who do this for a living focus their lasers first?

"Definitely the frame," says Johnson. "I like all the lines to make sense and work with each other. I get a bit obsessed with lines and curves matching, so modifying frames is a good job for me. When I figure out what tank I'm going to use, the frame usually changes with the tank choice."

"I decide whether the frame is as functional and beautiful as I'd like for it to be and whether the suspension is up to the tasks I might be throwing at it," Stulberg explains. "If not, then that is ditched too. We might simply start over almost altogether, as I did with the latest project we did [a Ducati J63]. We started with the motor and built everything else around it. We chose the desired handling, weight distribution, and ergonomics, then went from there. It seems that every bike I design is inspired by one piece. It can be a drum brake, an engine shape, a specific tire, or even a headlight. It really does vary from machine to machine."

A word of warning, though: the more you stare, the more you'll wonder whether or not you have the chops to hand-fabricate a seat pan or a battery box or a side cover. You'll wonder where to find a new ignition for that '68 Triumph or a horn for that '73 Honda. You'll wonder, in other words, where all the new parts come from.

John Ryland offers some advice: "I recommend choosing your mods to complement the motor and other elements that won't be going away. You might have a killer idea for using a certain tank only to realize it won't sit on the frame without riding on the valve covers. Be flexible with your design ideas and know when to move on from nonessential elements that fight with essential ones. There will be another bike. And another. You can use those ideas later."

No matter which parts emerge (and you're quite likely to learn a bunch of new skills in the process of figuring it out), every machine has its own something special that begs for

"Pick your favorite part of the bike and build everything else around it. It's okay to start with your least favorite part as well. Take the ugly, undesirable part and figure out how to either cover it, tweak it, or completely replace it with something that inspires you."

attention, begs to be first, begs to be the project's signature piece. It's up to you to notice which part keeps you awake at night. Just don't let it drag you down. It's a good thing. It's the call of the motorcycle. It's a future piece of art telling you what it wants. As Alan Stulberg so deftly puts it, "Pick your favorite part of the bike and build everything else around it. It's okay to start with your least favorite part as well. Take the ugly, undesirable part and figure out how to either cover it, tweak it, or completely replace it with something that inspires you."

Contrary to what you might think, this mental design process does not need to start with elaborate schematic drawings. When it comes to sketching out their ideas, our builders don't always go about it the way you'd expect.

Is it important to have an end goal in mind? Where do they start?

Jarrod DelPrado: "Definitely imagining and discussing the end result. We usually only do sketches toward the end so our painter can have specific ideas to work from."

John Ryland: "I Photoshop a lot of stuff these days, but it's out of necessity more than enjoyment. Clients need to know what they're getting. For you, it's okay for a lot of it to be a surprise. My only advice is . . . if you're Photoshopping, make sure you have your proportions correct, as you can't simply transform your tank to be an inch shorter and slightly wider."

Jared Johnson: "I have sketched, but it doesn't work that well for me. I'm more of a hands-on, in-the-moment designer, staring at it, walking around, putting different parts on, and, in some cases, mocking something up with cardboard and designing mentally."

Max Hazan: "Although I love to draw and paint, I usually don't sketch any bikes on paper. Instead, I weld the engine to the table at ride height, put the tires where they will go, and then sketch the bike full-scale on a sheet of wood behind the imaginary bike. It allows me to lay out the proportions and see how things will connect and interact with each other—length of the tank, seat height, shape of the frame, et cetera. I always draw the bike in my head. I can see it."

previous page:
Max Hazan says the Vintage Flat Trackers website (www.vft.org) is the secret treasure map for finding rare parts sold independently by their owners. If it's a rarity you need, email ads@vft.org, then cross your fingers.

FAMOUS LAST WORDS

Once you've processed all this at the end of the first day—once you've put the book away and turned off the lights and made the valiant effort to drift off to sleep, only to find yourself sitting on a shop stool in the middle of the night—you're going to feel overwhelmed.

Don't.

In the darkness of the shop, it's easy to trick yourself into thinking you can't do this. (It's also easy to convince yourself it'll be easier than it looks. Don't do that either.) The key is to take a lot of deep breaths and move one step at a time. Remember, all of these builders went through the same things you will.

Here's their hard-earned advice for how to get moving.

"I recommend working someplace where you can stand back and look at your bike," advises John Ryland. "Or better yet, sit and look at it, or lie on a sofa and look at it. Fall asleep looking at that sucker. Ideas will come to you as you take parts off and reveal the bones of the bike. For your first project, just think of how you want to use the bike, how much money you have to spend initially and how soon you want to be riding it. How much work do you plan to do yourself? These things should guide you as much or more than the vision of the bike in your head. To this day, I picture really cool-looking pieces made of interesting materials with subtle curves and unique finishes. Sometimes they work out and sometimes I realize right away that I'm not qualified to pull it off. When that happens, I move on to something that makes more sense but still has character. It's better than pulling your hair out over a small detail and delaying the build. Of course, that's my opinion. All of this stuff is."
Adam Ewing

"Build the bike one hundred percent for you and no one else," says Jarrod DelPrado. "If you love it, that's truly all that matters. But make sure you really build it the way you want it to be. Don't compromise and say to yourself, 'Close enough.' If a certain part didn't turn out the way you wanted it–even though it sucks to lose the productivity–scrap it and start over until you get it right. We've redone tons of paint jobs, tail sections, exhausts, et cetera because they were close, but not perfect." Here, Justin is at work on G2.
DP Customs, LLC

"I get a lot of emails asking for a parts list, or 'How did you do this or that?' for everything on my bikes," says Jared Johnson. "I appreciate the fact that people want to get the same end product like the bikes I build, but I feel like that would take away the fun for new builders. I would say, get in your shop and do what you think is good-looking and have fun with it. Design something new and different. In the end, you will have something that is completely yours."

Pierre Robichaud

"You have to make the design process a dialogue, not a monologue," Hazan advises new builders. "You have to listen to the parts. They will tell you what to build around them. It sounds a bit cheesy, but it's the truth. Go try and put a headlight on your bike. You'll find yourself moving it up and down until the bike tells you where it should go. Don't be scared to make mistakes or just dive in and make the cut. Ninety-nine percent of my bikes changed along the way during the process. Sometimes you can shoot yourself in the foot by overthinking the route to the finished product, or worry about making a mistake getting there. I have learned most of my skills through trial and error. Mistakes are inevitable and, in hindsight, they were never that bad. I say, just make the cut and get the ball rolling. With that said, the design process does come home with me and keep me up at night sometimes, so it's not all just from the hip."

David K. Browning / E3

"I think that I could find the beauty in almost any motorcycle," Stulberg claims. "Beauty can be found in its simplicity or, conversely, in its complexity. I think that one of the biggest strengths we have is our ability to distill the design of a machine so that even a complex machine can look simple and stripped down. It's truly an organic process that totally depends on the bike and project at hand. I often sketch things out either over a photograph or simply by looking at the machine. The basic outline is all that is needed."
Alan Stulberg

THE LINES

Something special happened in the late '70s besides the advent of cool Japanese motorcycles. In 1979 a woman named Betty Edwards penned a book called *Drawing on the Right Side of the Brain*. In it, she attempted to teach people with an inescapable inability to draw a straight line how to draw a straight line. Learning to draw, she said, depended on learning how to see. Not how to see things as you've always seen them—in three dimensions, with depth and perspective and texture and movement—but in two dimensions. Linear and still. As if the world had been flattened out on a sheet of paper like a chunk of Play-Doh you're about to carve your initials into with a pencil.

previous spread:
David K. Browning / E3

Max Hazan
Geoffrey McCarthy

NOT RULES, BUT LABELS

CUSTOM MOTORCYCLES | CLASSIC MOTORCYCLES | MOTORCYCLE GEAR | MOTORCYCLE WALLPAPERS | TOP 5s | SEARCH

« PREVIOUS YAMAHA SR250 'AG HOC' | TOBIAS GUCKEL'S KAWASAKI CAFE RACER NEXT »

HOW TO BUILD A CAFE RACER

By **Charlie Trelogan | On Cargo Collective**

I'm a car designer by trade: I spend my time working out how to make machinery look as good as it can. Designers are creative people by nature, so we crave the opportunity to be as free as possible in our work. We also have many parameters, tests and boundaries to refer to, to make sure we deliver the best possible 'product.'

These guidelines are just that—*guidelines*. Designing a café racer is as much about art as science, and each bike is different in its own way. It reflects the environment, the era and the owner of the bike. Yet there are things we can do to ensure that the result will look solid and professional.

I've been influenced by motorbike design for several years, and have built my own café racer. I based it on the same observations that I've sketched out here. Hopefully they'll inspire some fellow builders to invest time into the aesthetics of their project.

To illustrate my points, I'm using the Bike EXIF calendar cover star: Mateusz Stankiewicz's

SUBSCRIBE

Sign up for the Bike EXIF email. Get the latest customs in your inbox, as soon as we publish.

First Name

Last Name

Email Address

GET UPDATES VIA EMAIL >

This section is loosely based on a wildly popular BikeEXIF.com feature post, "How to Build a Café Racer," written by guest contributor Charlie Trelogan, a car designer by profession. Some of the terms used in this chapter came from him and represent the kinds of lines he sees in motorcycle design. It's vital to note, however, that they are not concepts necessarily meant to guide motorcycle design. Rather, they are terms used here to describe motorcycle design that has been done before. They're repurposed here to help you see bikes differently and to create a language you can use to make sense of the things you see when you look at a motorcycle.

It turns out that asking one of our builders to reduce his work to such basic constructs is kind of like asking Buddy Rich to play backbeat on a Jerry Lee Lewis song. While rock music mostly breaks down into a small collection of standardized structures (e.g., verse-chorus-verse-chorus-bridge-chorus-chorus), it's the bands that completely break those rules and ignore traditional structure that really stand out over time. Our builders are the Led Zeppelins and Tools of the motorcycle kingdom. Their lingo doesn't often map verbatim to these terms. In fact, there appears to be a lack of standardized lingo for the ideas in this chapter.

Even Charlie concedes that the terms in his Bike EXIF post were somewhat makeshift and devised mostly for the purpose of describing café racers. "The framework I outlined in the article," he says, "was absolutely only for café bikes and probably only really applies to café bikes of a certain age. Very little of it can carry over to the normal world of bike design, and even less to cars. Cars are more complex products, but there are some things that are the same but in another context. The bone line, for instance, is vital to the side profile of a car. It's the physical center of the vehicle, and where the horizon reflection will fall. We take weeks to get it right."

Codifying Trelogan's labels any further, he points out, could be limiting, and could sound too much like a set of rules where there really aren't any. So as with everything else you learn about design in this book, these are guidelines, meant to be ignored as much as they are considered.

"It kind of comes down to just having taste or not having taste," says Jarrod DelPrado, "or having flow or not having flow. It's no different than when you put your clothes on. You're probably not going to put on plaid shorts and a vertically striped shirt. For us, it's sort of a given, and sort of natural, that you want to flow the best you can when you can."

With all that in mind, some of Trelogan's terms do offer a nice framework you can use to break down the various shapes and patterns of a motorcycle's aesthetic while working on your own.

To that end, here are some things to think about while you're choosing your outfit for the road.

previous page:
Car designer Charlie Trelogan's guest post on Bike EXIF's blog post offers up some useful terms you can use to think about the design of your custom bike.

FOUNDATION LINE

This build by David Browning of E3 Motorcycles makes the perfect template for discussing lines, including the one shown here, *the foundation line.* *David K. Browning / E3*

The foundation line is the main line that runs from one end of the bike to the other (mostly) and serves as a sort of *equator* for the bike. It's what the brain sees as the most significant and defining line of the bike, even if almost nothing else goes along with it. It can be dead straight, curved like a tilde (~), slope up or down, dip down, arch up, or any combination of those things. Somewhere in that beautiful mess of parts, a dominant line sets the tone for everything else. It may not be blatant—it may even be disguised by other parts, such as side panels—but if you stand back and look at the dominant shape of a bike as if it's a stick figure, you should be able to see the imaginary line dictating its overall shape.

On a typical café racer, for example, the foundation line is straight, flat, and parallel to the ground. It's formed by the bottom edge of the fuel tank, and stretches straight to the back end of the frame hoop, which serves as the lower edge of the seat and tail. It's a classic, vintage look, and it creates the feeling of speed.

Compare the foundation line of a chopper and a café racer. The chopper has a fairly obvious *reclining* position. While this might be perfect for a highway ride, it's not going to conjure up any images of a prize-winning speed racer. The look of speed comes from matching the line of the bike to the line of the road it's sprinting down—long and flat. A café racer is a piece of paper resting on a tabletop.

Jared Johnson's signature Schwinn-style bikes, on the other hand, have a foundation line that curves from front to back, mimicking the roundness of many parts of a motorcycle engine and bringing to mind the narrow-waisted, wide-hipped, voluptuous, Bettie Page curves of the cars his dad kept around when he was a kid.

Regardless of how it looks, the idea here is to consider what kind of foundation line your donor bike has now, what kind of line you want, and then whether or not you can get there. If you're starting out with a bike whose natural lines already resonate with you, the easier thing to do by far is to leave it as is. If, however, you have a '67 Triumph and want to go hardtail bobber style, you'll need to chop the frame and make your own back half. And if you're going to do that, you can play with the line quite a bit.

Classified Moto's XV920-R6 features a stubby, short tail section that bends upward as it emerges from the fuel tank. This creates the feel of a bike that was built to go into battle. You're going to jump on it with a shotgun in hand and ride one-handed while taking aim at the beasts behind you.
Adam Ewing

"To me, action means fighting your way out of the city in the zombie apocalypse."

"I like it when a bike sits like its ready for action," says John Ryland. "To me, action means fighting your way out of the city in the zombie apocalypse. To others, it might mean dragging an elbow at Laguna Seca, or getting thirty feet of air. That's where the design choices come in."

Jared Johnson's Schwinn-inspired bikes have a clear, distinctive foundation line that serves as the basis for everything in their designs.
Jared Johnson

Jared Johnson's signature Schwinn-style bikes, on the other hand, have a foundation line that curves from front to back, mimicking the roundness of many parts of a motorcycle engine and bringing to mind the narrow-waisted, wide-hipped, voluptuous, Bettie Page curves of the cars his dad kept around when he was a kid.

At the time of this writing, Jared was frequently holding up his frame-in-progress to the Schwinn bicycle on his shop wall to compare their shapes. "I've been taking measurements off of it," he says. "Having the model here is really helpful."
Jared Johnson

Regardless of how it looks, the idea here is to consider what kind of foundation line your donor bike has now, what kind of line you want, and then whether or not you can get there. If you're starting out with a bike whose natural lines already resonate with you, the easier thing to do by far is to leave it as is. If, however, you have a '67 Triumph and want to go hardtail bobber style, you'll need to chop the frame and make your own back half. And if you're going to do that, you can play with the line quite a bit.

CUTOFF POINTS

At the back end, the seat sticks out just far enough to line up with the rear axle. At the front, the fender stops abruptly in line with the front axle. The matching cutoff points give the bike a harmonious, balanced look.
David K. Browning / E3

The front and back of a bike are not so much determined by its outermost edges, but rather by the points at which the bulk of its parts begin and end. These points can, at times, determine whether a bike looks silly and accidental or deliberate and considered.

One of the hallmark details of a bobber, for example, is its "bobbed" rear fender. Rather than stretching all the way around the top of the rear tire, the fender is chopped off near the horizontal center of the tire, ending directly above the axle. (Again, this is a generality. Custom motorcycle design is defined by its exceptions.) This creates a visual cutoff. On a bobber, nothing else significant goes past that point. When a bobber's taillight is mounted to the rear fender, for example, it is usually minimal and placed so that it doesn't stick out past the end of the fender. More often, though, the taillight is mounted to the side of the bike along with the license plate bracket, an approach that keeps them both nicely tucked away and not mucking up the clean look of the bobbed fender.

This Honda XL 600R keeps its cutoff points well behind the front axle and well ahead of the rear axle, giving it an aggressive and grungy look. It was built by Classified Moto for actress Katee Sackhoff and was featured on Velocity Channel's *Café Racer* TV series. *Adam Ewing*

A café racer's cutoff points, as illustrated in Trelogan's Bike EXIF post, are the center lines of both wheels. "Too much over the rear wheel," says the post, "will make the bike seem rear-heavy and poorly planned."

Jared Johnson says the same thing is true of the front: "I typically don't like anything going the past the front axle," he says. "I think it makes the bike look really long and a little bit sluggish. I like the bikes to have a little peppier attitude."

Jarrod DelPrado is looser on the subject. "I think it just depends on the bike itself," he says. "If we have a motorcycle where the tail section is kicking up from the ground, but it extends six inches past the rear axle, I don't really think you're going to get into a scenario of it looking silly. It can go the other way, where you can be too short. A lot of it is honestly just based on the motorcycle."

Max Hazan's rule seems more hard and fast. "For most of my bikes the cutoff point is at the axles," he says. "There is no particular reason other than that it felt right in that situation. For the modern bikes that I have built, the cutoff point never exceeds the overall length of the bike [tire end to tire end]. My shops have never been on the ground floor, and have always required maneuvering the bikes into a tight elevator. I have never worried about a tail or fender hitting something."

That said, a bike's cutoff points are very often crossed over, just not by anything significant. A license plate bracket might hang off the seat hoop. A taillight might curve up and out and create a little point of interest for the tail section. A front fender might have an ornament tacked to it like an old-school Rolls-Royce. But headlights and seat parts (like a sissy bar) and plastic will stay nicely in front of and behind the cutoff points.

Decisions regarding cutoff points contribute to the balance of a design (or its lack of balance). It's also vital to make all of these decisions look purposeful. In design, a slight inconsistency can appear to be a mistake rather than a decision. If two lines run parallel to each other, for example, and a third runs, say, a couple of degrees off from the first two, anyone looking at the trio will think it was the result of poor workmanship. If, however, the third line is a solid twenty degrees or so different, the brain sees it has having been done on purpose. This can be the difference between a bike that looks accidental and one that looks considered.

"If it's a curve or a flat line," says Alan Stulberg, "It has to be overt, and it has to be deliberate-looking. If it's going to be asymmetrical, it better look real intentional, or it's going to look like you were lazy and sloppy."

On a technical note, Alan Stulberg points out, it's important to consider whether or not the tire might hit the tail if the bike bottoms out during a hard ride. If a tail hoop ends directly over the rear tire, the tire can touch it. For this reason, tail sections of this length usually curve upward. Making the tail longer or shorter, to end before the tire or after it, can eliminate this problem. Another option is to increase the gap between the top of the tire and the frame's tail so that a bottomed-out tire still will not reach it.
Alan Stulberg

HEIGHT

This bike's seat height and total height aren't too far off from each other. The effect is a bike that looks flat and fast, as opposed to, say, a chopper, whose ape-hanger bars could be two feet taller than the tank, creating a more reclined look.
David K. Browning / E3

There are two types of height to consider in motorcycle design: the overall height of the bike, which affects the rider, and the bike's height limit, a design choice that refers to the tallest point of the bike's profile.

You need not adhere to any type of rule here—again, this is all about doing what you want—but it's worth understanding how these decisions affect the stance of a bike and its rider.

First, the overall height.

As Jarrod DelPrado mentioned previously, all of the bikes to come out of the DP Customs shop have had their rider's height taken into consideration. This isn't true for everyone.

"It's a gamble of what I think visually works," Alan Stulberg says. "In fact, the Hardley [a prior Revival Cycles project] is one of the biggest mistakes, but the owner is just fine with it. He's not very tall and has a short inseam, and the bike has a really high seat. He can only touch the ground with one foot, tiptoe. The seat is really high—not just the seat, but the suspension—but with the way we wanted to stance it, the way it looked right was to pull the motor up. We raised the center of gravity. But that bike is a blast to ride. It's probably the most fun of all of them."

If you're building a bike for yourself, it's entirely up to you how much you care about being able to flat-foot the thing at a stoplight. If you're building it for someone else, though, it might be a nice to ask the eventual owner how much they intend to ride the bike. If you're building on the hopes that some unnamed person will later buy the bike, it might also be wise to stay close to stock heights so you can appeal to more people. Build a bike with a thirty-inch seat height and you might be ruling out any potential buyers under five-foot-six.

Max Hazan isn't a small man. At six foot two, he has to consider how a shorter rider might look on one of his complicated creations. *Dikayl Rimmasch*

When it comes to height, Max Hazan focuses on the look of the bike as well—all our builders do—but part of this consideration for him is how the rider will look on top of it. (No motorcycle is complete without its rider, right?) To this end, he has a workaround for the tall guys. "Most sport bikes," he says, "can accommodate taller riders as well as shorter. However, most café- and bobber-style bikes tend to make taller riders look huge. I realized after building an XS650 bobber that if you build a bike with tight lines and have the seat right on the rear fender, it may look great, but you will dwarf it, which is why at six foot two, in order to get the right height while maintaining the bike's proportions, I just use bigger wheels and tires." (Max's version of a bigger wheel is significant; whereas most street bikes have a sub-twenty-inch wheel, Max will throw on a thirty-incher.)

Jared Johnson leaves his brat-style bikes mostly stock (besides his signature Schwinn-inspired builds, he regularly builds brat-style bikes for customers). But he still applies some control over this aspect of a design by trying to level out the height at the front and back, and even lowering the bike a bit overall. "I definitely cut out some spring in the front and lower the front end as much as possible," he says.

Um, and how do you do that, exactly?

"You just basically cut out some of the spring, and then you take the whole fork apart, then actually put a piece of tube in there (with the right diameter) to take up the space of the missing cut spring. If you just cut spring, then if you were to catch air, there'd be too much play in the front fork, and the fork would just drop out. I definitely do a lot of fork spring modification."

He swears this isn't as grueling a process as it might seem.

"It's not too bad," he says. "The main problem is that you don't really know how much to cut out. You start at two inches and cut the spring and then put it all back together and put it on the bike, let it sag, and then you're guessing, 'Okay I want it to be a little lower.' So you go in and cut another inch and then put it all back together, and put it back on the bike. It's time-consuming, but not really hard work."

While the height limit of Hazan's Ducati is consistent from the front end through the fuel tank, the tail pokes out a little above it, which accentuates the downward angle of the rear frame line. *David K. Browning / E3*

The other factor to consider in a bike's design is the height limit, the tallest point of the bike's profile. It's the vertical equivalent of the cutoff points described in the previous section. The height limit of a café racer, for example, is often the tallest point of the fuel tank. Generally, no other major part—the seat cowl, the headlight—is positioned higher than that point, though a fairing or tail fin can sometimes reach a touch higher. On many sport bikes, the tail fin is the highest point of the profile. This makes the bike seem as though it's pointing downward, and is also reminiscent of a NASCAR fin. On many custom bikes (especially vintage), however, there is no tail at all, so the fuel tank becomes the tallest point, and most everything else sits below it.

VISUAL WEIGHT

That triangle of open space behind the engine is a common sight in custom motorcycle design (it comes from removing the airbox, relocating the electronics, and crafting a more minimal seat). Even though the bike as a whole now looks lighter, the visual weight of the engine is even more pronounced.

"Visual weight" describes how a bike's density is distributed.

To understand this, consider which parts of a motorcycle allow you to see through to the other side of it. Spoke wheels, for example, have very little visual weight. They're light and airy. They feel transparent. The motor, on the other hand, is quite dense. There may be a few small open spaces around it, but the engine itself is a solid block of metal. The denser an area, the heavier its visual weight. Generally, the lightest areas of a bike—its most open areas—are at the front and back. The heaviest visual weight of many, if not most, bikes tends to occur just ahead of its horizontal center because this is where the engine sits. This is often true even when other areas of the bike are covered by plastic, such as on a modern sport bike or a scooter. While the plastic creates a solid area and, therefore, the illusion of substance, the engine's material—solid steel—always feels heavier. On other bikes, the weight can be more evenly distributed. In either case, the position of the weight affects its appearance and even the style it falls closest to.

"For me, the sportier bikes tend to have a more forward-positioned visual weight, while cruisers sit it at the rear," Hazan states.

Quite often in custom bike design, part of the builder's goal is to reduce a lot of visual weight (and actual weight) by removing parts, tucking away cables, and using smaller gauges and controls. There may not be much one can do about the placement of the engine, but even with that constraint there are some things you can do to balance out the visual weight if that's the look you prefer.

"Most of the time," says Alan Stulberg, "I don't like big. When all your visual weight is front-loaded, it just looks like it's got no balance at all to it."

To create the balance Stulberg is after, Revival's bikes often feature some version of a tail section that's more substantial and stylized than a mere seat hoop. It might be a thicker seat done up in a brighter color than the rest of the bike. It might be a racer-style seat cowl with a paint job that matches the fuel tank, accentuating the line started by the tank. It might mean blacking out the engine (powder-coating it black, or using one that was black in the first place) so that other parts of the bike draw more attention. (More on this in chapter 6.)

Between the minimal seat, the open spaces behind the motor, the tucked-away electronics, and the mag wheels, you can practically see right through this bike called G2 from DP Customs. All the visual weight is centered around the engine and tank. *DP Customs, LLC*

Another method for altering visual weight is to manage clutter. Clutter for a motorcycle usually comprises a lot of very necessary parts, but many of them can be moved around.

Jarrod DelPrado, again: "It's about asking how minimal you want it to look versus how bulky you want it to look. If you look at the frame of a Harley V-twin and the frame of a Sportster, which we're basing most of our bikes on, the motor and the oil tank and the battery all fit into really tight quarters. Really cramped up. When the bike is factory, there are all these wiring harnesses and boxes, and literally every nook and cranny is filled with something. It's like a wall. You can't see through it. We have definitely tried to reduce the visual weight of some of our bikes by opening them up as much as we can. We've had a couple of customers that don't prefer that. They want more of that bulky visual weight."

Regardless of the distribution of its visual mass, dense details can create the impression of actual weight. If you'd like your bike to look lighter, consider ways to open up spaces around the engine, get rid of plastic coverings, and reduce the mass of whatever is left.

"I try to hide everything," says Jared Johnson. "Basically, the engine and tank are the focal point of most motorcycles, in my opinion. The whole back area behind the carbs—between the carbs and the rear tire—I try to cut that out completely, and hide the battery as well as possible. I've actually taken the starter out and then tucked the battery in a battery box underneath the motorcycle where the starter is, really condensing everything to around the engine."

Jarrod DelPrado continues: "We prefer the open look. For our bike called The Player, the custom oil tank Justin made takes it from the factory location and puts it down low . . . removing the battery from that location and putting it into the tail section. Now you can see through a nice, beautiful, open space in the middle of the bike that, to us, looks more refreshing and makes the bike appear to be lighter weight. Certainly with bobbers it's the same thing. We're trying to be as open and light-looking as you possibly can be. That's important to us."

top image:
Jared Johnson does such a good job of minimizing the seat (note how thin it is), tank, and other elements of his signature bikes that the engine soaks up virtually all of the visual weight.
Jared Johnson

bottom image:
If you simply don't have much flexibility with your lines, you can work around by drawing attention to other things. Look at a few Classified bikes and you'll notice the frequent use of mesh-metal side covers. "Get creative with covers and shields," says Ryland, "to distract the eye from things that might conflict with the angles you are working with."
Adam Ewing

BONE LINE

You can practically draw a straight line from one end of this bike's bone line to the other. Note how the reflections form a straight line from front to back.
David K. Browning / E3

The bone line, as described in the beginning of this chapter by Charlie Trelogan, is the physical center of the vehicle, where the reflection will fall.

Trelogan was careful not to overapply this term to motorcycle design in general, but for the sake, once again, of building a language with which to talk and think about motorcycle design, this term might be useful. Looking at a bike and deciding whether or not you like it is one thing. Understanding how to create what you like is an altogether more complicated task, so forcing yourself to see a bike in terms of lines and shapes has value regardless of the accuracy of the terminology. Just like *Drawing on the Right Side of the Brain*, this is about learning to see differently.

Trelogan explains that the bone line is "the widest point of your bodywork." Whereas the other lines described here are meant to be seen like a line-art drawing, the bone line is 3–D, as the bodywork is the only part of the bike that has a reflection (unless your bike is covered with chrome). Bodywork includes the fuel tank, headlamp, side covers, tail section plastic, and seat cowl (whichever apply to your bike). Following Trelogan's minimal explanation, the bone line is defined by the widest points among these pieces.

To quickly get a sense of where the bone line falls, go outside and look around at the cars in your neighborhood. On a great many of them, you should be able to see a discernible line running from front to back that cuts through roughly the vertical center of the car door. Note how the reflections that show up on the car door (as well as in front of and behind it) are affected by this line.

Cool, right?

A clear bone line is one great effect of Revival Cycles' tendency to match the tail to the tank.
Alan Stulberg

This line is an effect of the car designers choosing which points of the bodywork will be the widest on each vehicle.

Now consider where this line falls on your donor bike. How does the headlamp line up with it? Is the headlamp's widest point different than the widest point of the fuel tank? Does this reflection point of the tail section line up with the fuel tank? If not, how far off is it?

If you replace the fuel tank of your donor bike with one from another model, you will change your bike's bone line. If you reposition the headlamp to higher or lower than its stock position, you will change it in another way.

One thing you can do, in other words, to define the look of your bike is to determine where you want these lines to occur, how they should line up, and *if* they should line up in the first place.

The 56 by DP Customs
DP Customs, LLC

PRIMARY LINES

This is not how Charlie Trelogan defined it in his Bike EXIF post, but for the sake of explanation, think of the primary lines as those that stand out the most. Here, it's the three matching lines formed by the frame and shocks and the exhaust, which stands out in contrast to them.

A bike's primary lines, for the sake of picking apart a design, are those that are most dominant after the foundation line. On any bike, there may be just a couple of them, or even just one. Even when there are multiple strong lines, they may not match. They may not have anything to do with each other. But if you do have strong lines on your bike, it can be good to consider how they match or mismatch other lines, and how all of the lines connect and relate to each other.

(Note that this is not how Charlie Trelogan defined primary lines in his Bike EXIF guest post. The term is being hijacked here to describe dominant lines in general.)

One candidate is the exhaust line. Exhausts come in a wildly varying array of styles, from perfectly straight to something more akin to a bendy straw poking out of a tropical drink on a Jamaican beach. However your exhaust line looks, matching the shapes of other parts to it can create a nice sense of togetherness. A dead-straight pipe can be accentuated by a dead-straight seat hoop that runs parallel to it. For a pipe with a couple of bends in it, you can fabricate parts (see more about this in the last chapter) that use the same curve radius. On a bike with cutoff points at or behind the axles, you could cut the pipe short to drive home the bike's feeling of compactness. Note, though, that in many cases, the exhaust runs along only one side of a bike, so whatever you do, your pattern-matching decisions may be less obvious on the naked side.

Another strong line is the one formed by the front bars of the frame. There may not be much that you can match to this line necessarily, but contrasting it with a line with a similar steepness that runs in another direction can make the two work together to create a more angular and severe look. The rake of the forks on many choppers, for instance, forms a wide upside-down V with the front line of the frame.

John Ryland uses this line in particular as a starting point for a lot of design work. "Classified Moto bikes are built using the stock frame steering tube," he explains. "This is a good choice for garage builders, by the way. Everything has to play off the fork angle, since we won't be changing it.

"Often, we use non-stock swing arms that convert the bike from dual shocks to a mono-shock. In doing so, we're going for a completely different look than the original frame provided. In these cases, it's not practical to maintain, say, parallel lines between the forks and subframe supports. So we try to choose subframe support angles that (a) don't conflict with other angles on the bike, and, when possible, (b) emphasize them."

The idea here is to work with what you have, complement it, and try to avoid screwing up the balance of other lines.

"It's kind of like the Hippocratic Oath: do no harm," Ryland continues. "There are plenty of cases where a visible bracket or support doesn't necessarily reinforce the design, but we try to make sure it doesn't hurt the design."

With the curve of the handlebar echoing the curve of the bike's foundation line, Johnson creates a sense of cohesiveness rarely seen in custom motorcycles. *Jared Johnson*

Le Mans IV by Revival Cycles
Alan Stulberg

When you're lucky, you can match a couple of the stronger lines of a bike to tie the design together even more. You can also turn a relatively insignificant line into a dominant line by matching it to one that's stronger.

Jared Johnson likes to add a detail or two like this to his bikes. One in particular is the kind of detail that wins fans. Remember that nice arch that serves as the foundation line for his Schwinn-style bikes? He mimics it with the shape of the handlebar. To make his handlebars, he bends a stick of metal pipe into the same arch he uses on the frame.

"It's maybe not a hundred percent the same arc," he says, "but I try to get it really close. If I just put straight flat bar on there—something like a drag bar—it wouldn't have the same flow to me."

It's the kind of detail you can never put your finger on when you look at a bike. Usually, a bike either looks cohesive or it doesn't. You either like or it you don't. If Johnson used a drag bar, it wouldn't look right. You wouldn't like it. With this handlebar detail in place, the whole thing comes together as if by magic.

Johnson, as you can see, isn't trying to match lines so much as he is trying to match curves, something that carries throughout his designs. "Down to even the fender mounts," he says. "I've sliced them off and put two other mounts on there. Just a straight piece of sheet metal with a couple of bolts in there isn't really going to cut it. I'll probably do some sort of an oval bracket for the fender. Just the little details that need to all coincide with the rest of the build."

When you stand back and look at the stick-figure version of your donor bike, what are the bare essential lines you need to draw it on a piece of paper? Whatever they are, you can do all kinds of things to work with them, whether that means matching them or contrasting them, and doing so can pull your design together to give it a vintage look, an aggressive look, or whatever look you might be going for.

SECONDARY/CONTRASTING LINES

You can think of "secondary lines" as supporting actors–they're a little less dominant and often contrast with a bike's primary lines. Exhaust can also fit into this category.
David K. Browning / E3

There are two things to understand about design that can help guide your use of secondary lines (and color, and material, and all kinds of other things, but those will come later).

First, symmetry is part of how we measure attractiveness. Fashion models, for example, tend to have symmetrical features, whereas the rest of us have one ear longer than the other, a drooping shoulder, or some other kind of ego-crushing flaw.

Second, when something stands out and catches our attention, it's usually because of the things around it. It's like the song, "One of these things is not like the other." A row of orange squares in a photo does nothing for us unless it has a bright green triangle somewhere in it.

Secondary lines are like the green triangle. To make something pop on a motorcycle, you first create symmetry, balance, and consistency (you'll notice this in other details as well, in later chapters), and then you break away from it. Two or more matching lines make a contrasting line stand out. The conflict creates drama. And drama is where your bike's stance really shines. With enough intention and willingness to experiment, you can end up with a symphony of lines that all work together to create small moments of beauty within a wall of details.

Alan Stulberg reflects on how Revival Cycles creates this contrast through their exhaust designs: "It's an absolute function, but what it ends up being—on this otherwise really straight, precise, crafted machine—is this contrast between rough, organic, curvaceous shapes, and these sharp, straight, specific proportions. I think design is always about contrast."

There are a lot of great things going for Revival Cycles' Hardley (a 2010 Harley-Davidson Sportster), not the least of which is the handcrafted exhaust, the lines of which work both with and perfectly against lines elsewhere on the bike. *Alan Stulberg*

On Jared Johnson's brat-style bikes, he does the work of consistency and contrast through angles. If the top frame tubes are flat and straight, for example, he sets the fender stays to ninety degrees from there and the license plate bracket to forty-five degrees. "If it's a straight line," he says, "and there are other straight lines, I try to square everything up."

Over at DP Customs, the DelPrado brothers take it case by case. "If you're going to be doing a hardtail, tight frame versus an uptick swing-arm frame, then let's say the primary line would be the backbone. And then the angle at which the tank is going to be following that backbone is kind of what I'd call primary."

Contrast also comes from a bike's secondary angles, the less significant lines that play off of the dominant ones. If the rear frame tubes match the angle of the forks, these are going to dominate the overall shape of the bike. Secondary angles, then, might be the rear frame tubes and the shocks. If the bike is mostly straight and flat and parallel to the ground, but the exhaust is at a thirty-degree angle, this is a secondary line you can use as an opportunity for matching other details, such as the angle at the back end of the seat hoop.

Jarrod DelPrado draws a distinction between the foundation line, which would be the top frame tube, and the bottom edge of the tank. Rather than forming a straight line, the two play off of each other.
DP Customs, LLC

FORK DISTANCE / RAKE

A narrower rake like this makes for better cornering. With the broad rake of a chopper, you might be better off avoiding corners entirely.
David K. Browning / E3

Rake is the angle of the front forks relative to a vertical line. It determines where the front wheel lands. It can significantly affect not only the look of your bike, but how it handles when you get it on the road. There's a big divide between a shallow fork rake and a broad one—much bigger than the foot or so difference in the position of the front wheel.

John Ryland explains: "At the extremes, more rake [think Easy Rider chopper] equals increased stability with decreased turn-in response. Less rake [forks approaching perpendicular to the ground] equals increased turn-in with a twitchy, unstable ride."

"Turn-in response," as Ryland refers to it, describes the feel of the bike as you physically turn the wheel while riding it. A lengthy rake might feel smooth and natural on a straight road, but around a tight corner, the wheel will seem to pull harder as you turn it, and this can mean adjusting the way you lean and handle a turn quite a bit. A tighter rake, on the other hand, increases the responsiveness and feels more controlled in turns and corners, which is why racing bikes have tight rakes.

Now think about how it affects the look.

Simply imagining a long rake conjures up the image of a '70s hardtail chopper. So if this isn't what you're after, your best bet is to find a donor bike with a more street- or sport-style rake. "Steeper forks will look more performance-oriented and will move the visual weight forward," Hazan explains. "I like to keep the forks as steep as possible, and keep the steering axis aimed close to the tire contact patch without making the bike unstable. I hate heavy steering and having to aggressively counter-steer a bike to make it turn—ride a Harley and then hop on a super moto and you will see how dramatically this affects handling—but responsiveness comes at the expense of stability. Think about how the bike will be ridden first, and then work back from that. Race-bike steering geometry is effortless, but can be twitchy."

Classified keeps their rakes mostly in stock form, but often lowers the front end a bit, resulting in an aggressive, forward-leaning stance. *Adam Ewing*

John Ryland explains what they do to enhance this aspect of a bike's design.

"Most of our bikes feature a modern USD fork swap," he says. "Typically, this lowers the front of the bike, thus decreasing the rake. If we ran a MotoGP team, we would meticulously refine the rake for our top rider, who could probably detect fractional degree changes. Instead, we want the handling to be safe and to suit the rider. When essentially raising or lowering the forks, there's a nice range of acceptable rake specs. We can tell if something looks awry during the build process without getting anal about degrees and formulas."

Regardless of what lines your bike has or what you hope to achieve from them through sheer force of will and manipulation, when this heady work is done, you can get on with the details of your bike's design.

This, for some, is where the real fun begins.

The Musket by Max Hazan
Sinuhe Xavier

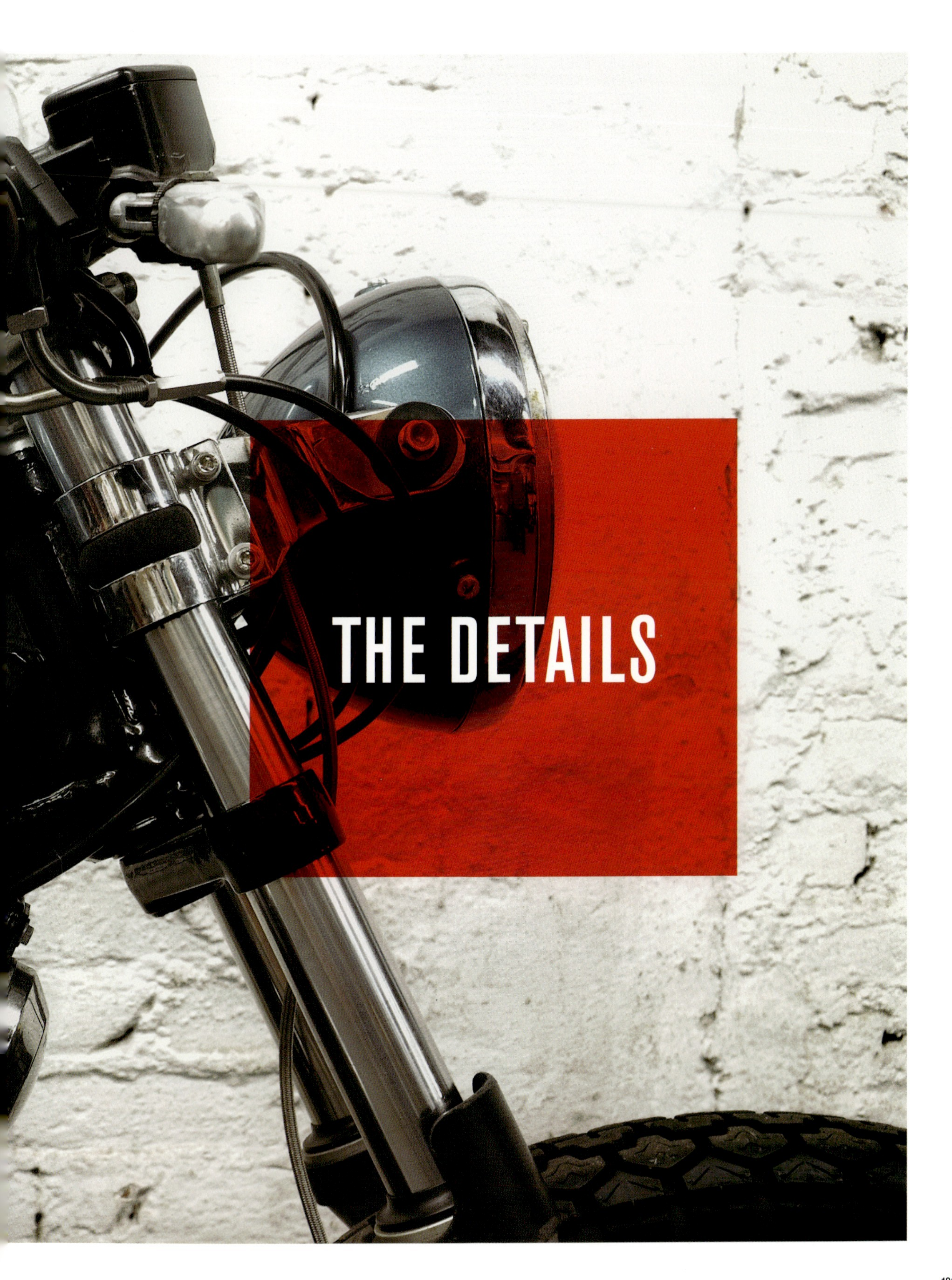

THE DETAILS

It happens any time you see a bike from a distance and then move closer. With each step, the machine grows more complicated. What looks like a whole package from twenty feet away becomes an orchestra of pieces, shapes, welds, scratches, nuts, bolts, plastic, and memories. The hundreds of little parts that make up a motorcycle come together to form its whole, and the moments that brought them together weave a tale unique to every one of them. Our machines are made of details, and their details are made of stories.

previous spread:
David K. Browning / E3

When you're lucky, you can hear a machine's life story directly from the person who experienced it. You can hear about how his cousin borrowed his bike once and got home without issue only to fall over at a complete stop in his driveway for no apparent reason. You can watch her eyes as she recounts the fluke moment she happened to meet someone with the exact same not-so-common motorcycle from the exact same year living just one town over. You can follow the line drawn by his fingers as he points at the pod filter he bought yesterday, the exhaust he replaced last week, the transmission he overhauled last summer.

When you see a bike sleeping in a parking space and its owner isn't around to share these stories, you can do something else: read the bike like a good mystery novel. Is that handlebar stock? Is that a motogadget speedometer? Is that scratch on the bar end from the time his cousin fell over in the driveway? I wonder if the tank was customized to fit that frame.

When you zoom back out, you realize it's these details that make up your impression of the bike. The feeling it evokes. Its raciness or its aggression or its relaxed attitude. Its balance. These details can even affect the way you perceive the bike's owner. They can tell you things about the person who throws his leg over the seat and tears off into the sunset. What he values. How she wants to feel while riding. How he sees himself.

This brings you back to your build.

While a bike's stance is pivotal to its gestalt, the true glory of any design is in its details. The wheels, the seat, how it sounds, the way the sun reflects off the handlebar, even the shocks. It all matters. It all has an impact on how you'll look at it, ride it, and talk about it. And every person who stops on the street to pick it apart in their minds will walk away feeling it was either a cohesive, thought-out design or a jumble of parts thrown together in someone's backyard. They may not be able to describe why they feel it's wrong when it is, but you'll see it when they walk away—that little shrug that says, "Meh. It's okay, I guess." (You'll get this response no matter what, of course, because people are hypercritical of work they didn't do themselves. Just do your best to ignore it.)

Whatever the reaction—yours or someone else's, considered or rude—this bike is going to earn plenty of them, and will tell its stories for years to come. Your stories.

Time to decide what you want those stories to be.

WHEELS AND TIRES

Believe it or not, wheels and tires are the perfect representation of the kinds of choices you'll need to make throughout your custom-build process. There's a host of questions involved in choosing them. Their answers, as they gradually decrease the possibilities, simultaneously increase the clarity of what your bike is going to become. Are you going to consider the rider or throw comfort to chance and go exclusively for looks? Are you going to honor a specific era of motorcycle design or follow your instincts and see what happens?

Regardless, the decisions you make here can steer your design in a lot of directions. Unless you've decided to do some major work to the frame, or to build your own, this is the first chance you have (in the context of this book, anyway) after choosing your donor bike to apply significant influence over how your bike will look at the end. Knowing what vibe you're after, if you're not just feeling it out moment by moment, can help you get there.

"The tires and the engine are usually the first two elements that I select," says Max Hazan. You may be limited to the engine that came with your donor bike, but everything else, budget allowing, is negotiable, and you're the only one doing the negotiating.

First, consider how you're going to be riding this bike.

One of the first questions the DelPrado brothers ask a customer when plotting out a new build is how that person plans to use it. Is it for sport or for Sunday strolls? (You considered these same ideas when looking over the lines of your bike in the previous chapter.) Do you want a nostalgic feel or something else?

Performance, looks, and style are all affected by the tires you choose for a bike. Think about how you'll be riding it, pit that against how you want it to look, and aim for something that works well for both considerations. *David K. Browning / E3*

The Hoosier by DP Customs didn't need a performance-focused tire, which allowed the brothers to run a stylish piece of rubber that looks like it came from a dragster. *DP Customs, LLC*

The customer behind the DelPrado bike called Two Lane, for example, lives in Indiana and planned to use the bike mostly on straight roads. He wanted to capture more of a dragster-type look. Hence, the brothers didn't need to worry about lean angles so much and were free to go with something more stylish. "We ended up using a drag-racing Hoosier front tire on the back of it," says Jarrod, "and a vintage Firestone style tire on the front."

A situation like this—where performance is less of a factor—can free up some possibilities. Fewer constraints can mean more options. You can even use the advantage, as Max Hazan does, to focus more on the specific look you're after. "If I am not basing the selection on performance, I like to use unique tires," he says. "I tend to gravitate toward race tires [old or new] with minimal tread. It seems to make the bike feel faster."

For all the complaining in blog comments about the trend toward minimal tires like the Firestone Deluxe Champions, something with a light and fast design, like this one from a Max Hazan build, can really add to the look of your bike.
David K. Browning / E3

MOTO
HONDA

"I love dual–sport tires," says John Ryland, "because they go back to our earlier rule of the bike looking 'ready for action.' There are those who will say an aggressive tire is a poor choice, but I believe the majority of riders can have just as much safe fun on an 80/20 dual–sport tire as on a sticky–sport tire."And then there's the matter of pairing. Like nouns and verbs in a sentence, like wines and meats at dinner, a wheel and tire combo can either complement or contrast not only each other, but the aesthetic of the rest of the bike. As Jared Johnson points out, it can even conjure up a strange image. "When I do the brat-style," he says, "I like everything to be fat. Typically, those tanks are pretty big. If you put 4.0 tires on a Honda CB500 twin or a 450 twin—they have big tanks, so I think it would look pretty weird with skinny tires. Kind of like how some big dudes have a big torso and then a small waist."

Wheel pairing also affects the overall angle of the bike. If you mean for your bike to have a fairly straight foundation line that runs parallel to the ground, a small front tire will affect the stance.

Also, a thicker tire creates a different impression than a thin one. A tire can either cause a passerby to feel a sudden itch to race or send him into a nostalgia coma over the Triumph he owned in high school. Whereas a thicker tire may imply substance, or heft, a thinner tire can imply speed. This pairs well with a café racer or another kind of sport bike where performance is a central concern. On a cruiser, however, it could look downright strange (of course, this could be what you're going for).

The size of the wheels matters as well.

"Normally," says Alan Stulberg, "I like them to be equal size, front to rear. It's classic and vintage looking. It looks balanced. It's really simple."

He notes that while seventeen-inch wheels might be an optimal size for racing, "We get a whole lot of performance out of an eighteen," and that "most bikes don't need a big, fat tire on them." This belief appears to stem from the fact that Stulberg is so focused on a bike's performance. "Turning and handling are better on thinner tires. The bike will be sharper and quicker in turns. Thicker tires don't do anything but give you more grip."

"I love dual–sport tires, "says John Ryland, "because they go back to our earlier rule of the bike looking 'ready for action.' There are those who will say an aggressive tire is a poor choice, but I believe the majority of riders can have just as much safe fun on an 80/20 dual–sport tire as on a sticky–sport tire."

When it comes to tires, Classified Moto doesn't screw around. Their beefy, knobby dual-sport tires create a meaty look that demands attention. This bad boy isn't going to take any crap from anyone.
Adam Ewing

DelPrado sums it up: "It has to do with the whole package. How you want to ride it, and then whether you're looking for a modern look with a vintage twist, all vintage, or something else."

Vintage style comes from referencing or mimicking all kinds of design choices made in those days, such as straight lines, spoke wheels, and minimal handlebar setups. A more modern sport style might require more chiseled and angled lines coupled with mag wheels. (Note that if you go with mag wheels, you are likely to end up with more visual weight around the wheels and less openness overall.)

For cost reasons, you may find it best to "run what you brung."

"I try to utilize whatever bike I buy," says Jared Johnson. "Basically, I try to buy bikes with good-looking wheels and a good-looking engine and take it from there."

But if you're up for it, you can break way out of the wheel-design box and work with a shop to make something custom.

"There's a huge difference in costs," warns DelPrado, "but it's a huge payoff in having a unique look." The effort, he says, comes back to them often in the form of compliments and questions about where they got those tires.

When a stock wheel just won't do, as long as you can afford it, you can find a local CNC craftsman to build out a custom wheel for you, like this one designed by the DelPrado brothers and whipped up in Arizona.
DP Customs, LLC

John Player Special

SHOCKS

Stock parts are meant to be broken and replaced with cooler ones.
David K. Browning / E3

It's quite possible you've not spent a day of your life thinking about motorcycle shocks. It turns out, though, that they are yet another detail that can affect not just the bounciness of your ride, but the look of your masterpiece.

Let's start with the height. If your donor bike's stock shocks are ten inches long and you want to lower the back end a bit, there are several options.

First, you can buy shorter shocks—eight inches, for example. If you do this, make sure beforehand that there is enough clearance between the rear tire and the tail section. Too little clearance and the tail can actually hit the tire on a hard

Jarrod DelPrado whips around a curve near the DP Customs shop in New River, Arizona, while testing the lean angle on the 56 bike.
DP Customs, LLC

bump, creating a shaky moment, if not something worse.

You can, of course, go the other way as well, by wrenching up some twelve-inch shocks, or even taller. One good reason to do this is for the raciness factor. "If we're going to do a café-style bike," says Jarrod DelPrado, "we ultimately like to raise the rear of the bike by an inch or so, maybe an inch and a half, and then we lower the front of the bike by about an inch. The real reason you want to raise bikes up—a lot of it in the rear too—is so that when you're leaning into a corner, you're ultimately raising the entire drivetrain, your feet and your peg, higher off the ground so you can get a better lean angle than you can when the bike is lower down. You're going to scrape a lot easier when the bike is lower."

If you want to keep the shocks that came with the donor bike, or throw on a set you already have lying around, you can try to locate a "lowering kit" that fits your bike. A lowering kit is a pair of matching brackets—one for each side of the rear tire—that mount to a bike's swing arm to create a new mounting location for the bottom shock mount, thereby altering the angle of the shocks slightly so that the total height of the tail is an inch or two lower than its stock height.

If you're as impatient as most of our builders, and have the chops, you can whip up your own shock mounts to change the height yourself. If you're already planning to powder-coat a bunch of parts later on anyway, a little bit of thick metal and some crafty shaping and welding will get you a modification that takes relatively little effort and has a big impact. This is what Jared Johnson does. He says, "If there's a huge gap between the fender and the tire—I'm talking five inches or something with stock shocks—then I'll get some lowered shocks in the rear. Sometimes, if the ones that came with the bike are really good, I will change the mounting points. I'll just weld up a new shock mount."

Next, let's talk about style.

A shock has certain aesthetic limitations by nature, but if you're trying to stay true to a genre or period of time, a mismatched set of shocks can earn you some "period incorrect" judgments (styles or parts incongruent with the time period the bike came from). If this doesn't matter to you, you'll have more options.

"If the decision is not performance based," says Max Hazan, "I tend to go minimal, which is why I build most of my own setups. If I was looking to go fast, Öhlins makes nice pieces, and they look almost as expensive as they are."

If the decision is performance based, you can still consider the look.

Jarrod DelPrado: "If we were going to be building a completely vintage-looking bike for a customer who said, 'I really want it to be all vintage; I want spoke wheels; I want it to look '60s throughout,' then we'd run some good shocks that are intentionally meant to look more subtle so they don't have anodizing on them or yellow springs or anything that would be a standout." For their 56 bike, however, they went with a racier look.
DP Customs, LLC

"If it's going to be a café racer–type bike," says DelPrado, "we normally want to go with a shock that looks aggressive and racy, one that has an external gas reservoir on it, like a Progressive 970, or like one of the upper-end Öhlins shocks."

However you do it, be mindful of the consequences, and do some research to make sure your suspension gets set up correctly.

John Ryland, with his penchant for aggressive attitudes, says, "On our swing-arm conversion bikes, we either run the stock shock—OEM Ducati, for example—or we use a Progressive Suspension 465 mono. We love them. We love Progressive. We buy a lot of their stuff and have a good relationship with them."

However you do it, be mindful of the consequences, and do some research to make sure your suspension gets set up correctly.

"There are tons of factors that apply to configuring and tuning suspension," Ryland says. "For someone looking to build their first café racer with their own two hands, I recommend first using common sense. If it looks sketchy, don't do it. If you must do it, get a good second opinion. Once you do it, test it carefully, and possibly wrap yourself in a Kevlar suit packed with several feet of fireproof foam rubber. Otherwise, stick to reasonable angles and proceed, as always, with the knowledge that if it breaks, it's going to hurt."

PARTS: TO BUILD OR BUY

If there weren't already enough challenges to a motorcycle build, one in particular will rise up to punch you in the face over and over again: the decision whether to buy parts or build them yourself. There are upsides and challenges in both directions.

The absolute first thing to consider is your safety. No one looks good while sliding down asphalt.

"For your first build," says John Ryland, "even though your budget might be limited, it's wise to ante up for certain pieces of the project. Anything that could fail catastrophically if you do it wrong should be a candidate for outsourcing. Anything that can be ruined if you get it wrong should as well."

Max Hazan agrees. "When I build my personal customs, I plan on making every part from scratch unless I coincidentally find something that works perfectly as is. I have gotten a bit carried away with my last few designs, and finding these parts is rare, but I jump at the idea of making less work for myself when it's available. The one factor that will deter me from using a part is if it compromises the aesthetic of the bike. I won't use a part out of convenience if it looks out of place."

Indeed, buying parts for a custom project looks like the obvious choice—that is, if you haven't done it before. By all appearances, it's the path that says, *Just pop on by the local speed shop or hop online, find what you need, and get that thing on the road! But as our builders have learned, it doesn't always work that way.*

First, there's the problem of finding the part. Aftermarket parts may have been available in every corner shop back in 1977 when your bike was manufactured, but finding them now can be like trying to order a rotary phone from the Apple Store. If your donor bike is even remotely old or obscure, it can be quite a challenge to locate anything that will work on it the first time you try. The odds of a speed shop having a part born to fit your bike are slim. And there are so many options online that it could take a week to arrive at a shaky purchasing decision, let alone a confident one. If you're looking for something other than a carb kit or foot pegs, you could find yourself looking for a long time. And the hunt can lead to a long list of sources.

One solid choice is eBay—the Motors section in particular. A breadth of parts is usually available, and the site asks sellers to specify which makes and models a part is compatible with during the listing process. This is optional, however, so while many listings do in fact provide this information, you're just as likely to have to guess. Buying used parts online is always a risk, but remember, as long as you pay a reasonable price for something in the first place, you can always resell it through eBay if it doesn't work out.

Craigslist (where you probably found the bike in the first place) can turn up the occasional gem as well. The search function leaves a bit to be desired, however, as sellers don't always use the same lingo you do. To get around this, use broader search terms. Don't search for a "carb kit for a 1982 Honda Nighthawk." Carb kits can be compatible with a range of models. Instead, search for "motorcycle carb kit." Generally, the less specific you are, the more search results you'll get.

Another nice trick is to use Google. Rather than use a website's search function, which is likely to be vastly inferior, head to Google and do a search for "site:somewebsite.com" followed by the item you want to find. For example, if you live in Austin, Texas, and want to search the local Craigslist for that carb kit, use "site:austin.craigslist.org motorcycle carb kit." (This trick should help you find your way through the shadows of any motorcycle parts site, for that matter.)

Lastly, here's a tip from Max Hazan: "I use VFT (VintageFlatTrack.org) for the rare stuff. They are a community of racers and enthusiasts who can usually track down anything you can think of."

If you're not having any luck online and your city has a motorcycle junkyard, plan to spend at least a couple of Saturdays there. Get to know the staff. Anyone who works in a motorcycle junkyard is bound to know a thing or two about bikes. They also have access to catalogs and can even introduce you to other customers who might know more than they do. "When I was getting into bikes, I liked the idea of modifying them visually and learning the basics to keep them running," Ryland says, "or at least being able to speak coherently about them to a mechanic. To me, the hobby of motorcycling—including wrenching on them for fun—doesn't equate to the need to know everything. In fact, I got a lot more out of the whole thing by seeking out people who know more than I know. Even though it cost me some money, I learned stuff that aren't in manuals and forums. Plus you meet a lot of interesting and sometimes likeminded folks when you include them in your project."

(Just remember to read the rules on your way in the door; junkyards don't always like it when you bring your own tools and just start tearing stuff apart. When you find something you want, ask a staff member to pull it and price it for you.)

Max Hazan explains where he finds parts (when he doesn't make them himself, that

You can always find a good selection of hard-to-find parts on eBay Motors, as well as the common ones you just don't want to bother trying to find in person through local shops. If you have the time to wait for delivery, hit the Web.

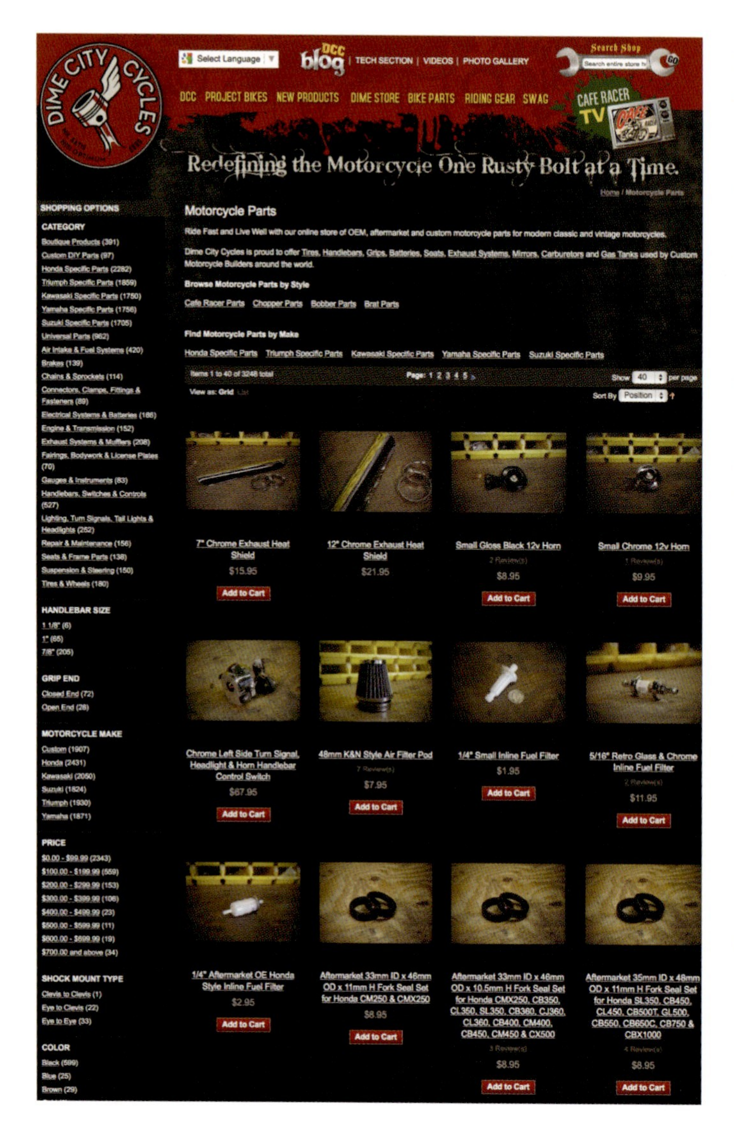

The "Bike Parts" section of the Dime City Cycles website is a great way to lose an afternoon. And a few hundred dollars.

is)."Anywhere. I troll junk stores, garage sales, and my collection of random parts from all of the donor bikes that I have cut up. eBay is still my biggest source of parts, albeit a bit less romantic."

If you build more than one bike, over timeyou're likely to end up with a big box (or a room full) of miscellaneous parts. Keep them. You never know when that extra fender can be cut in half, reshaped, painted, and thrown onto the back of a Kawasaki.

A major advantage to stocking up on trash parts is that they can leave you with something special later on. "We're always looking for something to recycle or repurpose because it's kind of an environmental karma thing," says Ryland, "but it's also a good idea from a creativity standpoint. Chances are, if a part looks good and is made inexpensively overseas, it's going to look like a great choice for a lot of people. Translation: Everyone is going to have the same thing on their bikes. But if you're using old transmission gears as foot pegs, that's a bit unique."

If you'd rather work with fresh-out-of-the-box parts, the list of possible sources gets infinitely longer. First, there are the motorcycle manufacturers, which sell parts through dealers and online through their own sites. This will work well for anything made within the last decade or so.

For vintage bikes, however, you're more likely to find new parts through independent shops that specialize in rolling their own or operating as resellers. These can be well-known motorcycle shops with websites or the one down the street from your house. Most repair shops sell at least some variety of common parts. In most cases, though, the more specific your needs, the less likely you are to find that exact part. If you absolutely must have an OEM (stock) part, you'll probably need to find it through the manufacturer.

But then, you might be sucking the fun out of your project. You have a lot more options if you get inventive. Rip out the old gauges. Throw out the handlebar. Ditch those oversized turn signals. Toss the plastic side covers. Order parts that will give your bike a new look—mini turn signals, café-style clip-on handlebars, a smaller ignition, bar-end mirrors. You name it, it can be modified. Go crazy.

If you go this route, Dime City Cycles (DCC) is a great option (see the "Bike Parts" section). Besides being well-known custom motorcycle builders, they also carry a good range of parts, including signals, grips, cable kits, rearsets (foot pegs designed to mount closer to the back end of the bike, common to café racers), all kinds of electronic parts, horns, fuel filters, tires, wheels, gauges—the list goes on for days. If you need it, they probably carry a version of it you can modify and weld into place. (A word of caution: Make sure anything electrical will work with your system. If you're not sure, talk to someone in sales support.)

Revival Cycles also sells parts online. Alan

and his team offer a much narrower selection than DCC, but you can be sure that anything Revival carries will be top-notch and as stylish as it gets. They even make some of it themselves using their in-house machine shop. If you're lucky enough to be in Austin, Texas, you can

Make sure your must-see list includes Bike EXIF (www.BikeEXIF.com), Pipeburn (www.pipeburn.com), Return of the Café Racers (www.returnofthe-caferacers.com), and The Bike Shed (thebikeshed.cc).

even stop by the shop's "lounge" to see some of the team's handiwork up close. As a serious bonus, there's a good chance that any Revival bike recently covered by one of the world's most popular motorcycle blogs will be sitting right inside the front door.

Speaking of blogs, there is arguably no better way to build up your sense of taste and style, short of going to a whole lot of motorcycle shows. On any given weekday, you can bet someone is posting a write-up about a fantastic new custom bike to come out of someone's garage, backyard, or apartment. Head over to your favorites, peruse their archives, subscribe to updates, and follow along on your choice of social networks. When you see something you like, make a note of what it was, which bike you saw it on, what you liked about it (its characteristics and how it affected the design), and the names of any particular parts that jumped out at you. Later on, when you're ready to get to work, you'll already have narrowed the world down to a much more reasonable number of possibilities. (The world is never bigger than when it has no constraints. Give yourself too many options and it can take you years to decide on anything.) Just pull up your list and start Googling.

Make sure your must-see list includes Bike EXIF (www.BikeEXIF.com), Pipeburn (www.pipeburn.com), Return of the Café Racers (www.returnofthecaferacers.com), and The Bike Shed (thebikeshed.cc).

Besides finding success through custom bike builds, Revival Cycles is also one of the world's most active resellers of motogadget parts.

If you aren't a regular reader of the Bike EXIF blog, you're missing out. Run by Chris Hunter, it's a deep repository of some of the best bike photography and build write-ups on the Web.

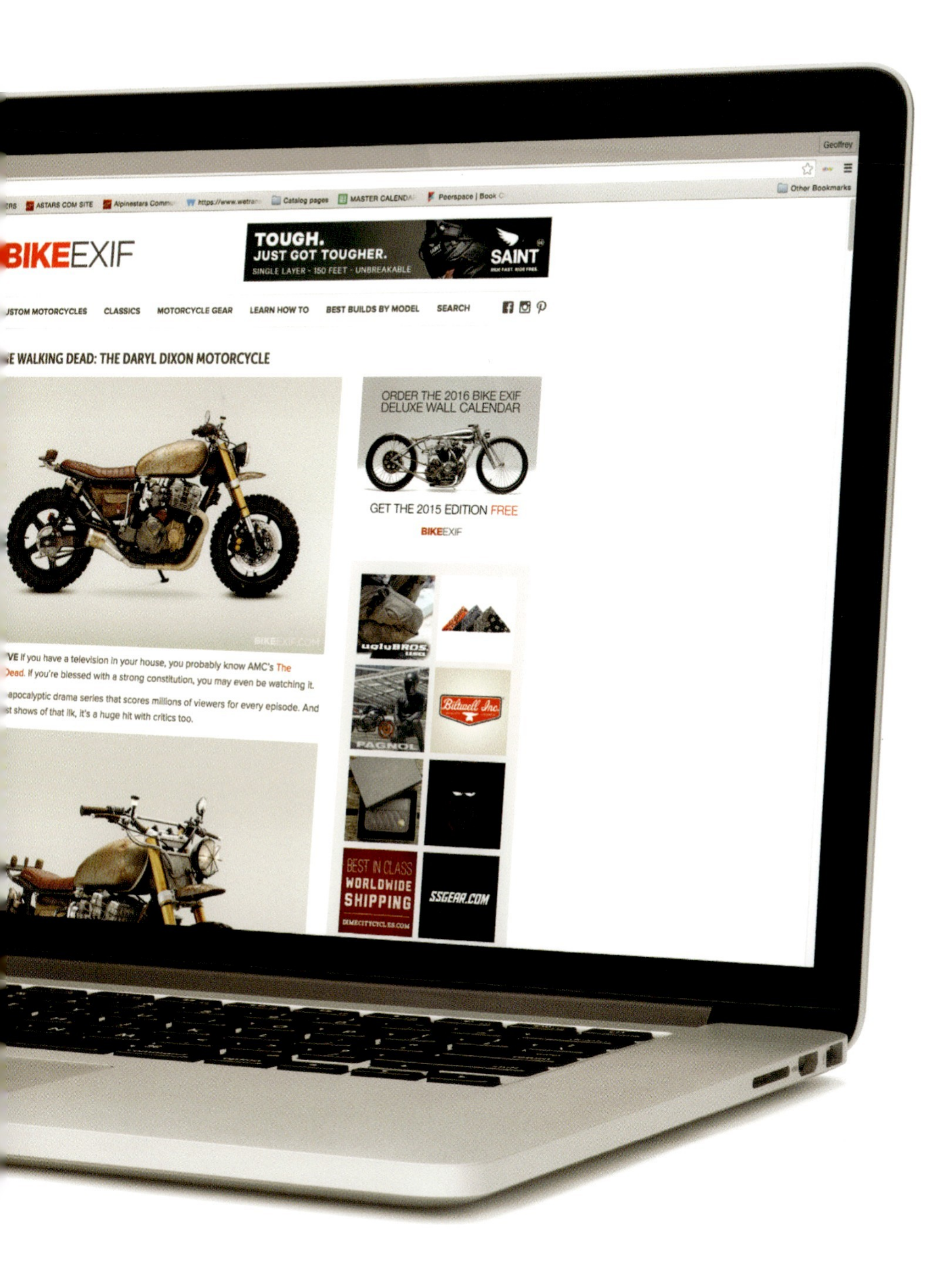

BIKEEXIF
TOUGH.
JUST GOT TOUGHER.
SINGLE LAYER · 150 FEET · UNBREAKABLE
SAINT
CLASSICS
MOTORCYCLE GEAR
LEARN HOW TO
BEST BUILDS BY MODEL
SEARCH
WALKING DEAD: THE DARYL DIXON MOTORCYCLE
ORDER THE 2016 BIKE EXIF
DELUXE WALL CALENDAR
GET THE 2015 EDITION FREE
BIKEEXIF
If you have a television in your house, you probably know AMC's The
If you're blessed with a strong constitution, you may even be watching it.
apocalyptic drama series that scores millions of viewers for every episode. And
shows of that ilk, it's a huge hit with critics too.
BEST IN CLASS
WORLDWIDE
SHIPPING
SSGEAR.COM

THE MYTH OF BOLT–ON

If you're not comfortable hacking away at metal to forge your own parts, it might be worth it to take a class or find a friend who has some useful skills (and tools). Because there's something really important they don't tell you in the manual.

"I can't remember where we heard it first," says Jarrod DelPrado, "but it's definitely true that there's one common thing about bolt-on parts, and that is: they don't. They never fit. There's no such thing."

This can be the effect of an imperfect manufacturing process or poor design or materials. A bike with a sufficient amount of wear and tear can also inhibit your ability to bolt even a well-made part in place. And if you modify the bike at all, you change what that shiny new part was meant to bolt onto.

John Ryland explains further: "Once you start chopping and cutting on a frame, you drastically limit the variety of parts that will bolt on and look good. So, often, even when you buy new parts, you'll need to modify them to work with your cool modded frame. Sometimes it's better to just start from scratch and build something that's made just for your application."

Unless you're doing a restoration, you may have the urge to mix and match parts from multiple motorcycles to create the beast in your vision. This can produce great results, but be prepared to do some bending and shaping and hammering to make it fit, even if the part is brand new and was intended as a bolt-on by its manufacturer.

"I've heard from a few people throughout the years that I'm working harder rather than smarter," says Jared Johnson. "But there's also something to that: I genuinely like making parts. Sometimes it should be easy, like on customer bikes that have a set plan and budget. Somehow, I still end up making most of the parts."
Pierre Robichaud

"From the very beginning," says Jarrod DelPrado, "you're like, 'Oh, well, where's that going to go?' And you have to find a place. I mean, you'd be shocked–there are so many modifications when we do a hardtail motorcycle. You have to weld on a place for the exhaust to mount. You have to weld on a place for the oil tank to mount. You have to weld on perches for the seat. You've got to weld on chain tensioners. Where's the coil going to mount?"
DP Customs, LLC

DP Customs has dealt with their fair share of this. "You're cutting. You're grinding. You're doing all this stuff," Jarrod DelPrado says. "Next thing you know, you've got many hours into a tank and then you've still got to do bodywork to it and repaint it and everything else, so even though it's a 'part that we bought,' it sure as heck is not a bolt-on."

Even more—let's call them "opportunities"—appear when you're simply trying to make things fit onto a bike that were never meant to go there according to the stock design. Bolt-on or not, any change in the placement of a part will stick you with the decision of how to mount it.

"The glamorous part of building custom bikes is chopping on the frame, swapping suspension components and tanks, and what have you."

Says John Ryland, "The glamorous part of building custom bikes is chopping on the frame, swapping suspension components and tanks, and what have you. But the reality is that when you're done with the bones of the bike, you'll be making a million tiny pieces that do things like hold the starter solenoid in

Before it can be built up, it must be torn down. A Classified Moto staffer takes the grinder to a donor bike.
Adam Ewing

place or keep the speedometer sensor five millimeters from the trigger magnet on the brake rotor. Those pieces need to be designed, made, protected [for example, painted or powder-coated] and installed. There are times during the build when you can make it easier on yourself by choosing to keep a tab or bracket instead of shaving them off in a blind fury on day one of the project."

The upside? You'll be telling people for years how difficult it was to move that ignition to the right side, and how you eventually figured it out. No bike tells a better story than the one you build yourself.

A round battery box is one of the many details Jared Johnson had to create himself for this bike.
Jared Johnson

MAKE IT ALL

If you're feeling ambitious, you can do a whole lot more than weld tabs in place. You can break out the heavy artillery and start handcrafting your own parts out of sheet metal and pipe. These are not everyday skills necessarily, but the intrinsic rewards of shaping your own inventions and then riding them around town can't be downplayed.

"Make it all!" says Max Hazan.

Hazan isn't screwing around like most of the rest of the world. The man makes virtually everything he needs in his own shop, drawing on a wealth of metalworking and woodworking experience to handcraft any piece he dreams up. But even with all that ambition, he's not the only one doing it.

One thing these builders have in common is that they all seem far more willing to make their own parts than to order them. (The builders, in fact, probably wouldn't be so notable without this urge.) Evidently, the tedium of Internet research is enough to do in a man for good. Not to mention the maddening wait for the UPS truck to pull up in front of the house (or shop). And, as they've mentioned, there is the problem of bolt-on products that don't bolt on. Like it or not, if you want to do something serious to that donor bike, you're going to need some shop chops and a tool or two.

The upside to this effort is that it invariably produces more unique results than off-the-shelf parts. No matter how stylish it might be, you'll always know that the off-the-shelf stuff is also being used on someone else's bike. If you want true individualism, there's perhaps no better way to get it.

If it was stock, it wouldn't be a Max Hazan bike. Here, Hazan works through one of the countless steps in the art of fabricating parts.
David K. Browning / E3

No stock part will ever look as cool as this custom piece by Max Hazan—a two-piece fuel tank with the bike's electronics hidden inside one half of it.
David K. Browning / E3

How much of this you do, of course, depends on your other constraints. If it's February 1 right now and you want this thing on the road in March, you're probably not going to be able to build all of your own parts unless quitting your job and sacrificing a few friendships is a real option for you. As Jared Johnson points out, it can take less time to make your own part than to find one that works and wait for it to be delivered, but there are also plenty of other things to do on a bike build, so spending all of your time fabricating parts will cut into the time you have to rebuild the engine. "You'll make mistakes and run into things that you didn't expect no matter what," says John Ryland, "but if you take a look at the whole project, you can decide where you want your time to be spent. Is your ultimate goal to build a fun bike that's reasonably reliable that you can ride all the time? Or is it important that you be able to tell those who care that you built everything yourself without help from anyone?"

If you care less about the timeline than the payoff of custom work, bust out the tools and get going.

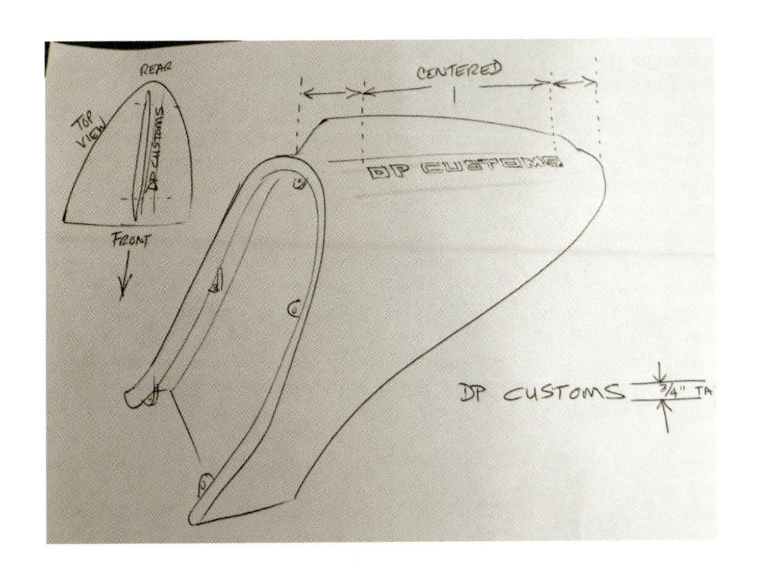

Jarrod and Justin DelPrado sketched out the tail for their Turbo Destroyer bike, fabricated it from scratch, primed it, and then had it powder-coated. There is no way they would have found a part like this on a website. *DP Customs, LLC*

THE LEARNING CURVE

If you don't already have the skills you need to modify a frame or whip up an ignition casing, you can start in a few different places.

Craigslist is a good option here as well. You can either post an ad to have someone do the work, find an ad by someone advertising those skills, or even enlist the site to find someone to teach you how to do it. There's really no limit to what you can learn, acquire, or sell through the site's free service. Oftentimes local companies advertise their services on the site (even when they have their own websites), making it the new-world version of the Yellow Pages. It doesn't take much to find nearby metalworking shops, classes, solo instructors, and people who are just willing to offer some advice in trade for some side cash.

To get the gear you need on the cheap—a useful alternative if you don't have much work to do and can bang it out in a day or two—you can often rent MIG welders and other tools from local shops. If you happen to be in Portland, Oregon, the home of Holiday Customs, for example, you can look up ADX, essentially the industrial version of a co-op that gives professional and hobbyist craftspeople access to CNC services, a full woodworking shop, and a full metalworking shop. A number of cities also have do-it-yourself auto repair shops that let you roll up and rent a bay for the afternoon. (To find them, try Googling the term "DIY auto repair shop.")

To enlist the help of a professional, try your local speed shop. In Phoenix, Arizona, for example, Ramjet Racing is fully staffed by people who have forgotten more than most people will probably ever know about building and customizing motorcycles. If it can be done to a motorcycle, Ramjet can do it.

Back to the Internet, there are a few tricks that will simply rock your world.

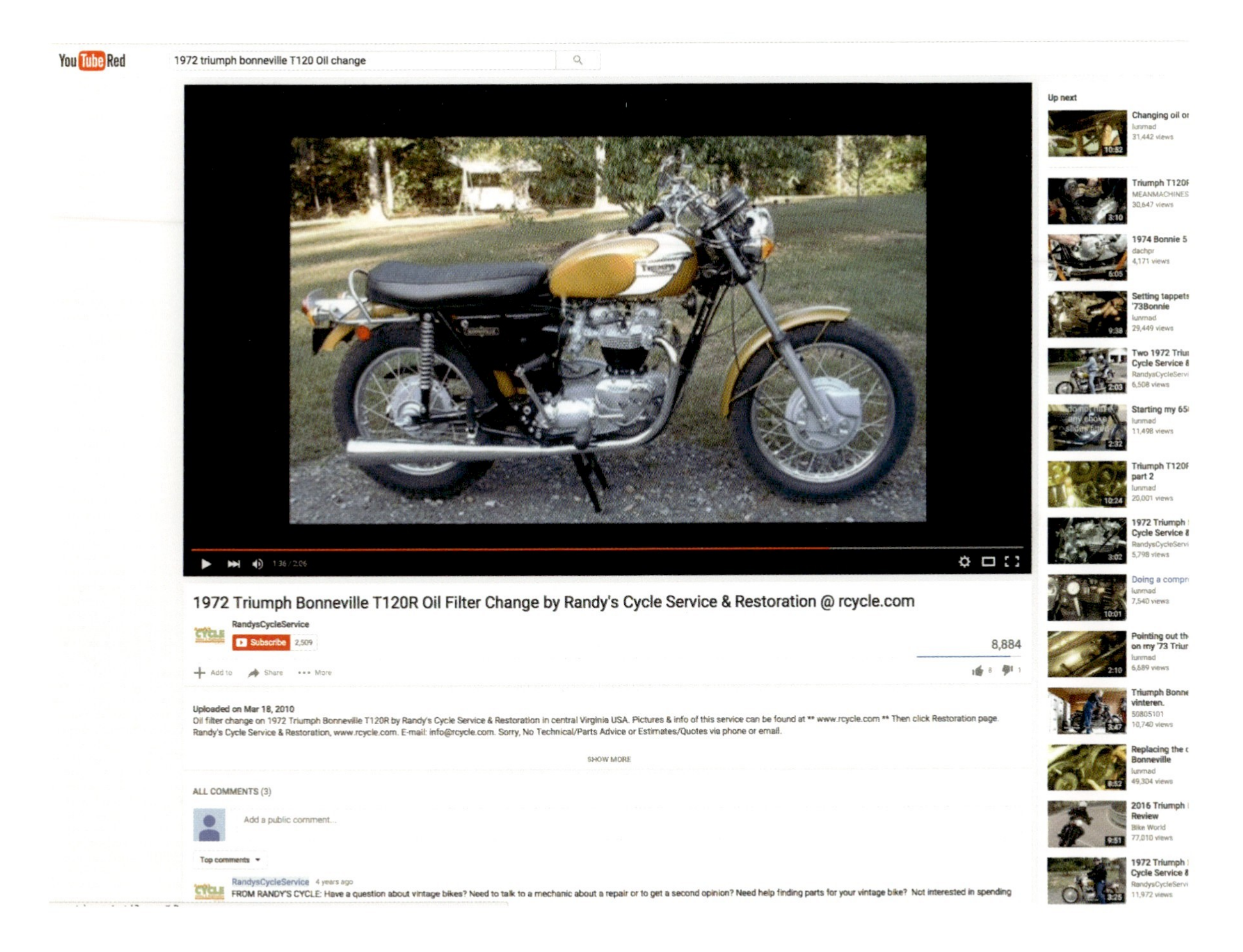

First, YouTube. This bastion of ridiculous viral videos also happens to have the universe's heftiest selection of amateur and professional how-to videos on virtually everything you can imagine doing to a motorcycle. You can learn to rebuild a carburetor. You can compare the growl of exhaust pipes. You can learn how to change the oil filter on a 1972 Triumph Bonneville. You can learn to cut steel, bend frame pipe, weld it all back together, and powder-coat it yourself. If it can be done to a motorcycle, you can learn to do it on YouTube.

When video instruction is too general, hit the forums. Whatever it is you need to know about—"how to fabricate a seat pan," for example—Google it. There will be forum posts on the subject written by dedicated weekend obsessives who felt inclined to share the bloody details of their own motorcycle projects. These write-ups often include step-by-step reviews of involved, months-long processes of stripping bikes down, cleaning up engines, creating fiberglass seat cowls with molds and cheap foam, and wiring up those fancy bullet-style turn signals from Dime City Cycles.

And if you have a question that hasn't been answered, forums are the place to ask it. Someone somewhere knows the answer. When people have this kind of expertise, or even just a solid passion for the subject, they hang around on forums and help out. It's one of the things the Internet does best. (That said, discussion sites are notoriously badly designed, so expect to be thoroughly frustrated while trying to use them. If you're not, well, then good for you.)

Learn the skills, get the tools together, and put on your garage clothes. Nothing beats the satisfaction of forging your own part, even if it's something as little as a tab to hold an off-the-shelf battery box in place. If you can sling a decent weld, you'll save time, money, and often considerable effort by making things yourself. And you'll have the pride of the job to carry around with you for years to come.

There is absolutely no shame in using YouTube videos to aid in your learning curve. The builders featured in this book have all done it, and it's one of the best resources on the planet for how-to lessons. Where else can you find someone to walk through a tutorial as many times as you can hit "Play"?

THE ENGINE

It was sometime in the early 2010s when we stopped noticing it. The mechanics of our machines had always been out there in the open to be looked at, considered, talked about, fidgeted with, and overhauled in any way we wanted. Then handheld technology came along and shifted the paradigm.

previous spread:
Alan Stulberg

next page:
Max Hazan's The Musket
Geoffrey McCarthy

At the turn of the century we bought up PDAs, then MP3 players, then cell phones small enough to fit in our pockets. Then came the smartphones, and with them, a new aesthetic we had not seen before: the perfectly sealed black box that left no hint of its assembly. It was perfect. Polished. There were no screws, no tabs, no moving parts at all. A screwdriver was useless other than to jam it into the face of the thing until it broke into a thousand pieces. When we inadvertently chucked one to the ground and shattered its screen, we took it to a store where we watched as an employee in an iconic T-shirt disappeared to a mysterious back room and reappeared with a brand-new unit.

At first, this alien object made some people question the nature of ownership. They wondered how a thing could be considered "ours" when we couldn't open it up, replace its insides, or turn it into something else of our own free will. They wondered how such an expensive and revered object could simultaneously be so disposable. *Got a broken volume button? Here's a new phone. Having trouble plugging in your earbuds? Here's another one. Battery not holding power? Yeah, we can't replace that. Take this one. Have a nice day!*

Over time, the vague discomfort wore off. The armchair philosophers who had raised perhaps reasonable objections joined the rest of us, so enamored by their devices that they simply stopped questioning the otherworldliness of their design. Collectively, we stopped noticing. If we must tear into things, we thought—if we must build and craft and shape and twist the objects that compose our reality—there are other things around to help us scratch that itch. We can repair vintage alarm clocks. We can handcraft leather wallets. We can refurbish old furniture. *We can build motorcycles.*

Indeed, there is one thing in our lives that has never and perhaps will never hide its mechanics so completely as to make them alien to us. One thing that is the perfect union of form and function, on which every part has a purpose, every piece contributes.

More than any vehicle on our great roads, and certainly more than any phone in existence today, the motorcycle still likes to show us how it works. It sticks its insides out there to be looked at, considered, talked about, fidgeted with, and overhauled in any way we want.

Admit it. It's a big part of what has always drawn you in.

There is arguably no more pivotal element to a motorcycle's design than its engine. Its sound. Its look. Its power. In a world where the drone of car engines is so common as to be unnoticeable, the rev of a motorcycle engine cuts through and demands to be noticed. When one whips by, it doesn't try to conceal itself. It screams for attention. And it makes no apologies for its primitive design.

For the next few pages or so, our builders talk about how engines affect their design choices and what can be done to maximize their three major effects: performance, looks, and sound.

THE DESIGN ANCHOR

"It is the central design factor," Max Hazan says about the engine he chooses for a custom bike. "It dictates the design of everything around it."

Remember, this is a man who literally bolts an engine onto a table and then hangs up a large piece of paper and sketches a motorcycle around it. When he says it's the core of the design, he's not kidding. For Hazan, it all starts with the motor. It's simply the one part of the bike you have no real control over. You can't relocate the cylinder head, or turn the engine on its side, or open up its visual density. The engine is what it is, and you simply have to deal with it.

"If your first project is like mine," says John Ryland, "you won't really be choosing the engine, but rather the bike, because it's cheap and available."

But if you haven't yet purchased your donor bike, there's a big broad world out there of engine design nuance, and you can pick and choose which one you're going to build your bike around. Each one has pros and cons, and each one has its own effect on a bike's design. Assuming you can get your hands on the engine (and/or donor bike) of your choice, it's largely a matter of your aesthetic style.

Here's a taste of the options.

No matter how cohesive the design around it, the engine is the core of any motorcycle. There's no getting anywhere without one, and it dominates the visual weight in a vast majority of cases.
David K. Browning / E3

Single: The single-cylinder engine is commonly known as a "thumper" for the sound it makes while running. It's light and nimble, but not so powerful. It's great for whipping around the neighborhood and is known for getting a hundred miles to the gallon. Every once in a while, someone even develops a bike show entirely to celebrate them. And since it's so mechanically simple, it could make an excellent choice for a first build.
1980 Honda Xr 500
Geoffrey McCarthy

The straight twin: Also called the parallel twin, this engine is common to British motorcycles like the Triumph (both old and new), and features a side-by-side cylinder setup with the exhaust pipes out front.
Lee Klancher

The flat twin: Here, the cylinders are horizontally opposed, sticking their noses way out into the airstream so they can be cooled by highway winds. The unusual design earns this engine another name: the "boxer." It's often seen on BMW and Ural motorcycles. If you use this type of engine on your own build, you should definitely go through a solid Google image search to get some ideas about how other builders have embraced this look and worked their bike designs around it.
Alan Stulberg

The V-twin: Also named for its cylinder arrangement, the V-twin's dual cylinders sit in a V shape around the crankshaft. It's used by Harley-Davidsons, Hondas, Ducatis, and several others. Ducati and Moto Guzzi set the two cylinders at a perfect ninety-degree angle from each other to reduce vibration, whereas Harley and others tend to use a shallower angle, allowing more vibration, which can be tiring for your arms on a long ride.
DP Customs, LLC

The inline-triple: This engine is set up transversely so that its cylinders form a line along the side of the engine rather than the front. Vintage inline triples are hard to come by, but very cool. *Triumph Motorcycle*

The inline-four: This was the heartbeat of the legendary Honda CB750 "superbike" and is still very common today. The four cylinders are usually lined up longitudinally across the front of the engine. Bikes slinging this type of engine commonly feature two-into-one exhaust pipes, which either run on the same side or with one on each side of the bike.
David K. Browning / E3

When you build bikes that
look this good, you can use
whatever engine you like.
Keep at it, Max.
David K. Browning / E3

The list of engine types goes on, but these are some of the most dominant players on the scene.

Our builders all have preferences and their own reasons for them. If you plan to use an engine other than the one that comes with your donor bike (or if you plan to build a bike around a motor), here's what some of them have to say about the decision.

"[I stick to] one- and two-cylinder engines," says Hazan. "They are much more aesthetically pleasing, sound better, and are narrow. I generally like to use carbureted, air-cooled engines, as there is much less peripheral equipment needed to make them run. I like the engines to look like they have just enough parts to work, and have all of the parts be interesting to look at. That said, I do have a sweet spot for modern supermoto/flat-track/MX engines. They are simple, light, and have a ton of power for their displacement—and they sound amazing."

"Sometimes customers have a preference for a certain engine configuration—'big V-twin' is a popular request—but they are more often drawn to the bike design than the power plant."

For Jared Johnson, the style of his bikes and performance dictate his engine preference. "I wouldn't want to build this style of a bike and then throw in a curved engine," he explains. "It's got to be round. There are a lot of square heads out there. In my opinion, there are a lot of bad-looking motors. Also, four-cylinders are not really my jam. They're way too wide and physically underpowered for having four cylinders."

While John Ryland agrees with Hazan regarding singles, he's more open-minded than Johnson about fours. He says, "Sometimes customers have a preference for a certain engine configuration—'big V-twin' is a popular request—but they are more often drawn to the bike design than the power plant. These days, I really like the simplicity of the big Honda singles [XR650L and XL600R, for example]. You can get a lot of usable performance out of them in a nice lightweight package. We've also had good luck with the later-model CB750 Nighthawk inline fours. Good solid power plant with plenty of available spares.

Ryland continues, "V-twins look super beefy from the side and skinny from the front. Inline fours look big from the front and not so much from the side. Singles are compact overall and look similar to parallel twins from the side."

(Like most of our builders, Classified Moto tends to use a variety of platforms for their custom projects.)

As for DP Customs, Jarrod DelPrado states, "Primarily we've done Harley-Davidsons and we like to use a 1200cc motor, and we go through every motor." (That said, they've also converted a Triumph Bonneville into a DP Customs masterpiece and intend to expand into a variety of other platforms.)

Whatever engine you end up with, use it for what it brings. Design with it rather than against it. If you try too hard to counteract the natural aesthetic of the engine, you could burn a lot of time trying to make things work that just won't.

Not to say you can't pull it off. John Ryland reflects on his first experience:

"My first project was a Yamaha XS850 Special. It looked like a cruiser with a big triple, and I wanted it to be more of a rat café bike. It was a good exercise in forcing something to conform to my vision, which is often the task at hand. It was the first bike I put USD forks on, and that really overcame the other drawbacks—shaft drive, little sixteen-inch cruiser rear wheel, cruiserish tank. The motor was neither here nor there visually, but it sounded amazing. [It was] my favorite part about the bike, by far. The more I rode it, the more I wanted the bike to look like it sounded, so ultimately it informed the visual character. That was kind of an interesting case."

The Rusi was a plain old Triumph Bonneville until DP Customs took it through their process. The switch from standard to metric tools probably wasn't the hardest part.
DP Customs, LLC

following page:
This Royal Enfield engine serves as the foundation for another Hazan Motorworks build.
Max Hazan

PERFORMANCE

One of the big questions you'll want to ask yourself when choosing an engine or donor bike is what kind of ride you want later on. If you plan to spend your weekends at any open-track day you can find, you'll want to start with something powerful and make it better. If you'll be happy just having something to get you to the corner and back, or if you're just learning to ride (or will be using the bike to teach someone else), a thumper (single-cylinder) in good running order might be perfect.

If you just want a little more power than you started with, a new air intake and some pipe wrap are all it takes to boost the horsepower a touch. *Robert Hoekman Jr*

Beyond preference or circumstance, with a little know-how, you can tweak your motor to get it closer to what you want. A good air intake and new pipes, for example, can add substantial horsepower to a bike—even one with a tiny engine. New carbs can help too. If you don't know how to do these things, again, YouTube is the ultimate source for how-to videos. You should have no problem finding someone to show you how to get your horse running stronger and faster. And remember, the guys at the local speed shop always know quite a lot. Go make friends. Your bike will thank you for it.

Our builders have all gone through their fair share of engine work. They've come away with a variety of insights, including what to look for when you get started.

"I think that I have had every single problem with donor bikes possible," says Max Hazan. "Since I can't open the bike up [when buying it], I first look at the owner and his garage, as I mentioned. It's usually a good indicator of how the bike was cared for."

Max's advice on how to inspect an engine prior to buying it is quoted here for accuracy's sake:

1. *Voltage test the charging system. It's an easy test, and you'll eliminate having to trace the issue through the whole chain of electrical.*

2. *Run the bike and check for oil. The outer covers are no biggie, but take a look at the cases and between the cooling fins where the head gasket is. Resealing the cases and/or milling the mating surfaces on the heads/cylinders may not be worth it in many cases.*

3. *Look at what comes out of the pipes. Usually black smoke can be remedied relatively easily. White smoke may take a little more time and money.*

As mentioned earlier in this book, be sure to prove that you can start the bike from a cold state. Any time you go to see a donor bike, touch the engine to make sure it's cold before starting it up.

A motor sits waiting in Max Hazan's shop, somewhere between teardown and assembly.
David K. Browning / E3

A warm engine means the owner started it up before you got there, possibly to conceal issues with the electrical system or other problems.

Once you have the engine, it's time to give it new life as best you can. Even if you have no intention of trying to boost its performance, a little work now can keep your used motor in better condition for a long time to come.

Jared Johnson doesn't even bother with the current state of the engine; he just gets to work on cleaning it up and getting it into good shape. "A lot of the bikes that I get, I don't fire them up. I just start building them right away rather than finish the build and maybe clean the exterior of the motor or something. It's a standard thing to go through the top end—at least to make sure the transmission is looking good. Then go through the whole top end and give it all new gaskets."

But, he warns, be sure to do this first. "If I put an engine back into a brand-spanking-new bike, with all the paint and everything, I don't really want to have to take the engine out two or three times." In other words, don't get the whole bike together and then decide to work on the engine. You could scratch up the nice new paint job (though, if you powder-coat rather than paint, this is less of a risk, as our builders will point out in the next chapter).

Beyond maximized working order comes the question of whether or not to beef things up. John Ryland tends to keep things simple for Classified Moto bikes. "Most often, we rebuild the motors to stock specs—or to whatever overbore is necessary—rather than going for max performance," he says. "We want to do everything we can to minimize problems down the road."

The last part of that statement, according to Jared Johnson, is important. The act of souping up an engine can actually cause some problems of its own. "I've never been big into upgrading horsepower," he says. "I always feel like if you want to have a faster bike, just buy a faster bike. I suppose it's something that my dad has always told me: You start tweaking with getting more horsepower, you're going to have problems with the drivetrain. It was built to handle 650cc. If you bump it up, you're just going to have more trouble down the road."

Jared Johnson doesn't worry about boosting the performance of his bikes. To the contrary, he worries about what might happen down the road if he does.
Pierre Robichaud

It's an important point. Of course, if performance is your thing, the potential for future problems probably isn't going to stop you. You can always fix it later, right? Just make sure you have a contingency plan.

"I have to use a lot of self-restraint on most projects," Hazan says. "My natural tendency is to want to push an engine as far as possible, but this often comes at the expense of reliability, durability, and practicality. On my personal bikes, I tend to get carried away and pay little mind to practicality when I am the one that will be riding it. However, where I have learned the most about engine work is through trial and error ruining my own engines—blown up on lean nitrous, bent connecting rods on too much turbo boost, et cetera. AAA is a must."

"I have to use a lot of self-restraint on most projects," Hazan says. "My natural tendency is to want to push an engine as far as possible, but this often comes at the expense of reliability, durability, and practicality."

At Revival Cycles, the goal is always to leave the bike in better shape than it started. The team prides itself on making improvements to an engine that even the manufacturer might concede is better than what it intended. They'll go so far as to build a standalone computer specifically for testing a fuel-injected bike, throw it on the dyno, and make it run the best it possibly can. Rather than merely clean a carburetor and throw it back on the bike, they'll build a new one. They'll optimize a bike every way they can.

"Well, it may not be absolutely optimized," says Alan Stulberg, "because there are always improvements that can be made with time and money, but it's better than it was. Yeah, we're not going to make it worse."

And that's exactly right. With time and money. If you have some of both, you can do whatever you want with your engine. Just don't make it worse.

Revival Cycles/Henne BMW Landspeeder engine gets the black glove treatment. *Alan Stulberg*

SHINE IT UP

Poke around the custom bike blogs for a minute and you'll notice the vast majority of them seem to be running sparkly new engines straight out of a factory. This is deceptive, of course. Most of the custom builds you see on these sites are based on old engines that have been shined up and given new life.

Grease is good. Dirt is better. But if you're going to go through the trouble of rebuilding a motorcycle and customizing every inch to your liking, it seems that many builders agree there's a case for starting at zero. After all, your frame, tank, mirrors, tail section, and more are all going to get some love through this process. Do you really want to assemble them around an engine that looks like it just clawed its way out of the junkyard?

You may appreciate the personality that comes from a bike that's been through some rides and has collected the grime to prove it, but if you're planning to sell the bike, your buyer might want to collect his own grime. Shine it up.

There aren't really any rules for how to go about it—start anywhere you want, be as thorough as you can, and take as long as you need. But our builders do have some tips.

"Take the whole thing apart and go piece by piece," say Max Hazan. "The results will be worth it."

Alan Stulberg agrees. "We'll take a motor apart and port and polish the heads, use new cams and new cylinders, and balance the crankshaft. We'll go the full distance with a motor."

No doubt the act will take some time, but the effect will feel like you've gotten a brand-new motor to go along with your brand-new everything else.

The gear you need to achieve this effect ranges quite a bit, from the very cheap to the relatively expensive.

Max Hazan's favorite is the 3M Scotch-Brite buffing wheel. "It's less abrasive and invasive than blasting," he says, noting that media gets everywhere and it's harsh on aluminum. A buffing wheel goes onto your angle grinder or bench grinder.

It won't always be hard, but it can be, and there are all kinds of approaches to getting at it.

"I spent a lot of time on my personal bike," says Jared Johnson. "Now, after five years riding it around and having oil all over it, I like it better. But I'd rather sell a bike to someone with a nice fresh-looking motor, even though I don't mind some old oil and some grime on there."
Jared Johnson

"Some of them are clean enough to where I can go over them with Brillo Pad," says Jared Johnson, "or just even engine degreaser and a pressure washer. Some of the engines come out looking real nice. This one in the last bike that I built, I had to bring to a machine shop where they boiled the cylinders and the head. It's basically like an engine dishwasher. The next day they deep-blast the exterior, the engine, and then put it back in the dishwasher. It comes out like raw aluminum again. I'm going to try to go to a radiator shop. I heard they dip engine parts in acid and it cleans it off perfect, and doesn't hit the aluminum like deep blasting would. It's really hard to get in all the nooks and cranny with a wire brush or Brillo Pad kind of thing. I do like it to look nice."

Alan Stulberg's team at Revival Cycles digs deep, and even digs things out to replace them. "We use full medical-grade ultrasonic," says Stulberg (an ultrasonic is a machine used in hospitals to clean surgical tools to sterilized perfection). "It's heated to clean all parts in the carburetor. And we typically replace all brass parts and whatnot. We don't just take the jets that were in it. We get a jet kit and decide which are the right jets and where the needle should be."

The DelPrado brothers lean heavily on S100, a line of cleaning products designed specifically for motorcycles. "It's like a mild degreaser that doesn't jack the finish up." As you can see from their bikes, it seems to be working. They do a thorough baseline cleaning, and then take all the parts off of the motor and clean them individually. "Then you've got perfect access to get in and to detail it so it truly looks the way it looked when it rolled off the factory floor," Jarrod explains.

Over at Classified Moto, John Ryland and the team make use of abrasive wheels. "When the motor is being rebuilt, we'll blast certain pieces, but I don't like the idea of blast media getting in places where it's not supposed to be."

And what about changing the color of your engine?

"It's just like anything else: the better you prep, the better it will look."

"There are good fuel- and heat-resistant paints on the market that we use," says Ryland. "It's just like anything else: the better you prep, the better it will look. If you're using engine paint or other heat-resistant paint, definitely read the directions. And I've found that using a heat gun between coats to "cure" it seems to help give the paint a durable finish. Also, no matter how fuel- and heat-resistant a paint claims to be, brake fluid will eat it. So will gas if you leave it long enough. We happen to like raw aluminum, though, so many of our motors are not painted and we don't have to worry about spills."

previous page:
"There's a lot of work that goes into the cleaning of the bike," says Jarrod DelPrado, "because we want the motor to look one hundred percent new, so we obviously break the motor down. We take the motor out of the frame, and I have a cart that I put the motor on–a special little cleaning zone in the back of the shop that I take it to."
DP Customs, LLC

SOUND

So you're into motorcycles.

Because of this, there are a few kinds of sounds distinct enough to grab your attention on the road. One is the high rev of a sport bike. Another is the growling, standoffish plume of anger coming out of the typical Harley-Davidson exhaust pipe. Another is probably the deep, throaty hum of a souped-up vintage Ford or Chevy pickup truck.

Maybe you're not a gear head. Maybe you never considered yourself the type to care about the sound of an engine. Once you become obsessed with motorcycles, you notice. Before long, your head turns like it's on a swivel whenever your hear a motorcycle race past you, and it's capable of pulling you away from any conversation just long enough to glimpse the lovely beast producing it.

You probably want your bike to be one of those. One of the head-turners. One of the conversation-interrupters. But stock exhausts can sound filtered and weak. Small engines can sound, well, small. Even when a bike has decent exhaust already, it can sound just a little left or right of how you want it to sound. It wouldn't turn your head when it rode past. It's gonna need some work.

"Sound is a huge factor riding and owning a bike," says John Ryland. "Customers sometimes have requests as to the noise level—some want quieter, some want loud. I like low and raspy. I also like the helicopter thump of a big single at idle. I'm not smart enough to tune exhausts like a musical instrument. But if we fire up a bike and it sounds lame, we go to work changing stuff to rectify the issue."

There are some tricks to getting what you want. The first step is knowing what you want.

"We're always kind of looking for a throaty sound that's loud but not obnoxious," says Alan Stulberg. "I wouldn't want it pointing in my ear. That's about it. You just pull the exhaust that's more open than a factory thing, and when it's a good motor it's going to sound good." But then, Revival is capable of things like this–a winding roller coaster of an exhaust pipe, handcrafted in their shop in Austin, Texas. *Alan Stulberg*

Do you want throaty? Tinny? Do you actually like the stock exhaust on a modern Triumph Bonneville? (They tend to sound a bit like a sewing machine, but, hey, maybe you think of it as "stealthy.") What you get depends on what you do with the exhaust pipes. Bent, straight, or curvy and wild, the way a pipe is shaped and how it's made affect the sound and performance of a motorcycle.

Straight pipes are a beloved option for a lot of builders, presumably because of their no-nonsense looks and fierce attitude. A minimal bike—say, a bobber that's long and low to the ground—with a set of straight pipes will slide around town looking like it's perpetually ready for a bar fight with a featherweight. A bagger with a straight pipe, on the other hand, will be commanding and bold. Either way, you'll have to deal with the sound that comes out as a result of this choice.

"I have been lucky that almost every bike that I have made so far has sounded great even with straight pipes," says Max Hazan, who adds, "One thing that I have found is that when you run an exhaust with no baffle/silencer [or mechanical back pressure], having several bends in the pipe will diffuse the sound so you don't have an ear-piercing, hollow exhaust note [as well as running the idle circuit a tad rich]. Choosing a setup for the particular engine is important for sound too. If you put the same straight pipes on a Shovelhead and a Honda Shadow, you would have one that sounds amazing and the other like a broken Honda Civic."

Jared Johnson has a process worth noting here. "I'll know right away if it sounds too tinny a lot of times—like with the shorter exhaust, it sounds airy and tinny. Not very deep. On pretty much all of the bikes with somewhat of a straight exhaust, I'll go in and put in little check valves. Half-moons of metal inside the exhaust."

Basically, he explains, you stick a cutoff wheel onto your angle grinder and then you slice into the exhaust pipe at a point about half the length of the straight section. Then you cut a small piece of sheet metal—smaller than the inside diameter of the pipe—stick it into the slot you just cut, and weld it in place. Then you repeat this on the other side of the pipe at a different location farther toward the back.

"So the exhaust has to do a little S-turn to get out," Johnson explains. "That creates back pressure. Helps the bike run better." It also tones down the tinniness of the exhaust note, he points out. Then "fire it back up and see if that helped out with the airy." If not, add more. "I'll probably do three or four, depending," he says.

People in the know might question this choice later on if they don't notice your slivers of sheet metal. If you want to guarantee they don't notice your homemade check valves, you can smooth out and buff the weld to create the appearance of a seamless pipe.

DP Customs is always playing with the exhaust sounds of their bikes. "We definitely like to enhance the sound," says Jarrod DelPrado.

In this particular example, Jared Johnson created an indent in the pipe to do two things at once: create back pressure and provide a home for the kick-starter.
Robert Hoekman Jr

"We think stock Harleys don't sound good, so we are certainly always trying to improve that sound when we're building pipes and, if anything, we like to make them have more of a higher pitch rip."

To this end, they've even gone so far as to standardize a box-shaped pipe they now sell individually. What else they do, as always for the gents from Arizona, goes back to the customer.

"Do we go with the wild level of loud, or mild?" Jarrod asks. "Sometimes we'll do a complete tuning so the pipe is wide open and the bike just rips and sounds insane. Other times, we'll do bikes for a customer who has a more subtle personality in general, and then we'll do an internal baffle inside and quiet it down. We always want to try to change the sound of the bike with regard to how loud it is or isn't going to be. It just depends on how crazy the customer is."

The trouble with all this is that you're not really going to know the effect of your pipe design on the sound of the bike until you get it all together and light it up for the first time. But even if it comes out a little thinner or weaker than you prefer, there can be an upside to how your design improves the look of the bike.

"Do we go with the wild level of loud, or mild?" Jarrod asks. "Sometimes we'll do a complete tuning so the pipe is wide open and the bike just rips and sounds insane."

Jarrod DelPrado again: "Justin always does the exhaust last, and that just finishes the lines, because he'll tuck it in tight and put the lines where they need to be."

It's a nice reminder: every detail of a bike affects the others.

THE FINISH

There is momentum at the ends of things.

The last few pages of a novel notch up your heart rate in anticipation of whatever it is that's about to be imprinted on your psyche. The end of a foot race reenergizes you for the last push to the finish line. The end of a project leaves you buzzing, eager to see the final product. When you can see the end, you can make it to the end.

So it is with building a motorcycle. If you're down to assembly and grips and seat leather, you're down to the last bits of excitement before you wipe the whole thing down with a rag and start it up.

previous spread:
Alan Stulberg

It may take a little while afterward to fine-tune everything and eke out the best performance you can from the shiny new beast in your garage, but rest assured, you'll get there. In the meantime, you can still have some fun.

There are just a few things to take care of first. And they might require some patience. Unless you're going raw and rusty, you'll need to prep everything for paint and powder-coating. Assuming you're not doing it yourself, you'll need to find someone who does this kind of work (ideally, sometime long before you need them back for assembly). And unless someone's going to send you a gift basket full of accessories, you'll need to find or design, make or buy the grips, seat, mirrors, turn signals, and every other tiny detail needed to finish off your bike. You need to choose colors. Styles. Stripes. Finishes.

There are just a few things to take care of first. And they might require some patience. Unless you're going raw and rusty, you'll need to prep everything for paint and powder-coating.

For a long time to come, these are the details that will draw in passersby who happen to notice your bike in the parking lot. Since they won't be able to judge the performance of the engine while it's sitting still, it's these pieces that will have them ogling your work of art with the curiosity of a child on his first trip to the zoo. They're also likely the pieces that will stand out the most to you every time you open the garage and wheel out into the driveway. A raw-steel tank in the sunlight. A red leather seat at sunset. Your handmade grips when you slide on your helmet and pull away from the coffee shop. The unscratched perfection of the frame, and pipes—even the gas cap.

You'll feel proud of what you did. Then you'll feel like you could have done better. Then you'll take a deep breath and feel proud again.

Don't panic over all the choices you are about to make. By the time you get close enough to see this part of the process, you'll be moving faster by the minute. Adrenaline will get you through. There is momentum at the ends of things.

PREP

If you're reading this book, odds are you've prepped and painted something at least once in your lifetime. Something. But have you done it to metal?

To plastic?

Before you can layer on the new, you need to strip away the old. That ancient paint that's chipping away on the frame needs to come off before you can make things new again. But how does that happen? Is it sandblasting?

"Sometimes," says Max Hazan. "I usually go for that as a last resort with engine parts. It gets everywhere. I go for the 3M buffing wheel first."

Sandblasting is the process of removing paint and surface rust by shooting an abrasive media at metal under high compression through an air gun. It may sound like a quick fix, but perhaps only if cleaning your shop is more fun to you than working on your new bike. There are simpler methods.

"We use paint stripper to—uhhh—strip parts," says John Ryland. "I'm surprised more people don't seem to know about this. It gives a nice smooth finish. We sandblast some parts for a textured finish. We also use various abrasive wheels."

Abrasive wheels and buffing wheels can be gotten on the cheap at just about any home improvement store or ordered online. Slap one on your angle grinder and buff away.

While a frame has far fewer nooks and crannies than an engine, it's large and composed entirely of pipe, which can make stripping an awkward process. "I'll go over it with a wire wheel on a grinder," says Jared Johnson, "or a Brillo Pad, and get it all down to raw metal."

He points out that it's not as brutal and time-consuming as it might appear, and it might actually save you more time than you spend. "By the time I've loaded up the frame and brought it over to the sand blaster and paid probably fifty dollars or something for ten minutes of sand blasting, picked up the frame, and driven it back to the shop, I figure doing it myself actually [is faster]."

He says he spent just fifteen minutes or so on one recent build when he was going for a relatively rough look. "It's just around forty years of old paint on there and a lot of chips and rusty areas," he says. "The wire wheel just plows through it."

For something more polished, you might expect to put a little more elbow into it.

If your bike has plastic parts you're planning to keep, common paint stripper can work there as well, but it might require a high-pressure washer to push things along. If you're like most custom bike builders, though, you won't be keeping the plastic. Alternatively, you can do what John Ryland does and create new side covers out of metal mesh. (Taking this idea to its furthest edge, DP Customs once worked with a company in Phoenix to invent an air intake using a 3–D printer. It's one heck of a way to avoid dealing with old paint.) Or do what bobber builders do and just get rid of the side covers entirely.

They may have avoided paint prep for this project, but DP Customs also had to invent an air intake. This one was made using a 3–D printer. *Robert Hoekman Jr*

COLOR

Once the old paint has been stripped away, you're free to think about color. But once again, it's a bigger subject than you might think. It's not merely about picking your favorite and moving on. Besides choosing which color to use, you need to decide how to use it.

Color serves several purposes. It sets a tone. It draws the eye. It creates or continues a theme by contributing to the feeling the rest of the bike tries to create. It tells people where to look. It can even redistribute visual weight. If thinking about it in this way feels a bit abstract or foreign, here are a few notions to help you view color the same way you now view the lines of a motorcycle.

Color serves several purposes. It sets a tone. It draws the eye. It creates or continues a theme by contributing to the feeling the rest of the bike tries to create.

First, take a look at which pieces of your motorcycle feature color. Odds are, your list includes the fuel tank, the side covers (if your bike has any), and the tail section (again, if applicable). If you're stretching it, you may also think of the choice between black or silver that was used for the handlebar and mirrors.

Did you consider the frame? What about the engine? And the wheels? The headlight casing? Turn signals? The seat? The grips? Literally all of these parts can be turned bright pink if you feel so inclined.

With any luck, that last sentence has you cringing but also now seeing your bike as a collection of distinct areas of color.

The frame might be black. The turn signals might be chrome (something you may now be reconsidering). If you have mag wheels, they might be black. Your engine is likely either silver or gray, or maybe blacked out.

Doesn't it seem, in fact, like anything that isn't painted is either silver or black? It doesn't have to be this way. It's your bike now. And you can do what you want.

Next, look at how the color is used. Where it's located. How it's balanced. If the engine, handlebar (and all its bolted-on parts), and seat are all black, perhaps you see an obtuse V shape when you step back and follow the line those pieces work together to create. If the engine is raw metal, but the rear wheel, seat, and front wheel are black, maybe you see an upside-down V.

What if every part of the bike was black, but the frame was white? Or bright blue? What would you see then?

It's amazing how much the lines of your bike can change when you start using color to change them. A drop here, a swipe there, and suddenly you're in a territory the manufacturer almost certainly didn't venture into. Their designers were constrained by the need to appeal to the widest market possible. You're not. You're a market of one.

Alan Stulberg says the Revival Cycles team gets a lot of compliments on the colors of its bikes. And indeed, they do use some unconventional colors. But the compliments probably have as much to do with how the color is used. While Stulberg says he generally makes color decisions at the last minute, quickly and without regrets, the success of these choices very much depends on all the work the team does before that decision is made.

opposite page top:
The Hardley by Revival Cycles draws all of your attention to its raw-steel tank and exhaust–well, and also to those two bands of raw steel at the top of the cylinder heads with the round air intake in between. Did you notice that they form a wide "M" shape? And that it perfectly matches the bottom line of the tank? And did you notice how these things together place all of the visual weight of the bike at the top, toward the front?
Alan Stulberg

opposite page bottom:
Revival's 1980 Moto Morini looks like a fairly straightforward bike. But did you notice how the seat color works with the wheels to create a long, wide-open look? If this bike had a small black seat, it would look fast and light, built for speed. With its long leather seat, however, it looks relaxed and casual. The tank and headlight on this one almost disappear.
Alan Stulberg

this page top:
On Revival's Ducati Pantah 650, you're probably only looking at one thing: the tank. Though the engine is bright silver, the creamy tank color draws your eye right to it and makes everything else just fade away.
Alan Stulberg

The John Player Special pays tribute to one of the DelPrado brothers' favorite pastimes: car racing.
DP Customs, LLC

DP Customs uses color for a very different reason: to pay homage to their biggest influences. The brothers are lifelong NASCAR fans, and as such, they often like to mimic the designs used on famous cars and by famous drivers. Rather than using it balance out visual weight or to draw the eye someplace, they're trying to conjure up images of race day at the track.

Color can also be used for a much simpler purpose: it can give a bike gravity or lightness or some other effect. "When I get the wheels powder-coated black," says Jared Johnson, "it makes the bike look real solid to me. A little more utilitarian. For the last five vehicles, the first thing I did was go outside with some rattle-can spray paint and spray those wheels black."

Even if you're going with a raw–steel look or retaining the bike's hard-earned rust and wear, you can protect it, even bring it out. "Anything we want to have a bare or patina finish we nickel-plate, shoot with clear powder, or both," says John Ryland. "Sometimes, we'll shoot clear-coat [automotive paint] on them if the part can't be heated in the oven, for example. Typically, we powder-coat frames and other pieces with a textured satin finish."

Ryland notes, however, that it's important to treat this process with a high degree of scrutiny or you can end up with parts you still need to work on yourself. "Choose a powder-coater who preps things nicely so you don't have to spend hours cleaning bearing seats, threads and the like. Generally, you get what you pay for."

Max Hazan agrees: "It is rare to find people who will treat your project the way that you will as the builder."

Builder beware. Ask for samples, inspect the work with your own two eyes, and, whenever possible, avoid sacrificing quality for cost savings. This bike should last you a long time. It's worth it to cough up the dough.

Insights from DP Customs

Jarrod DelPrado has a lot to say on the subject of powder-coating and painting, the people who do this work for a living, and the effort of doing it yourself. Presumably this is because, in days gone by, he was the brother responsible for suiting up and spending hot summer afternoons inside of the tiny paint booth at their shop in New River, Arizona. (If you've never been to Arizona in the summer, imagine baking inside of a tin box turned up to about 130 degrees.)

His lessons learned are now your words of warning.

ON POWDER-COATING:

"Powder-coat versus paint, for us, just comes down to the strength of one over the other. Early on, we painted our motorcycle frames ourselves so we could control the color and the finish and everything. But a painted motorcycle frame, even if prepped perfectly, is far more fragile than powder-coat. When you're putting the motor back in, if you accidentally barely bump it, you're going to chip it. Powder-coat is just such a strong coating. It didn't take us too many bikes to realize that. Ever since, for years now, we've been getting all of our frames powder-coated. The same evolution happened with the wheels and on the motors. You want to have really good heat durability over the long haul, so you're not going to paint something that could degrade in time and chip more easily."

ON POWDER-COATERS:

"A lot of people out there are trying to get into the powder-coating business. You can get some powder and a sand blaster. You'll see dudes that are actually powder-coating stuff in their own ovens, in their own kitchen. If you want perfection, I think, you've got to find a really good pro.

"Finding a powder-coater with that level of quality can be difficult, I think, because a lot of them tend to see more high-production, in-out deals. We were lucky enough to meet our powder-coater early on. They know the level of customers we have and all that.

"Over the years we've learned from each other. Nine times out of ten, they come out perfect. Sometimes they don't. Debris can get in there. It's just fact of life. As long as they're cool about making it right, that's good."

If you go with powder-coating (and you probably should), take the time to find a professional. There's nothing more beautiful than a part that's just come back from the shop.
DP Customs, LLC

The DelPrado brothers have begun sketching out their designs as part of a collaboration with a local painter. *DP Customs, LLC*

ON PAINT:

"The really key design element of the motorcycle is the paint job for the 'tins'—the gas tank, the fairing, the tail section. We never powder-coat anything like that because that's the true artistic finish of the bike and powder-coat is just not going to get that for you. Those we always paint."

ON PAINTERS:

"Paint and body work is a very lengthy part of doing a bike.

"For our bike called The Player, for example, I think that from beginning to end, we went in and out of the paint booth thirteen or fourteen times. It starts with your two coats of primer in the beginning, and then the multiple base coats of paint, and then multiple coats of clear-coating, then finishing with matte finish and clear. Going in and out of the booth every time is a ton of work. You're suiting up, you're getting on your mask and your suit. You're getting all your mixings perfect and dialed in just right, and cleaning the guns every time, and on and on and on. There's a ton of work mixed in with the fact that you're trying to do everything else on the bike.

"Early on, it was a big source of pride for us that we painted our own motorcycles in addition to all the other work, but thankfully the word got out and our build times started stacking up. We had more and more customers and bikes waiting in line. Painting is super time-consuming. Pretty much no other builder out there that we're aware of does their own paint. Then you realize when it just comes to the economy of working with a professional painter. All they do is paint, whereas we only paint every couple of months during that phase of a bike. It just made sense to contract somebody else to do our paint for us.

"It really has been an awesome thing,

because there are so many things that can go wrong during all the different steps of paint and bodywork. We have millions of nightmare stories we could tell you over beers one night about how many little things have gone wrong during paint jobs. When you're super detail-oriented, you can't let a tiny little speck on a gas tank go. You have to fix it. It's tough work to do.

"Eventually, we found a gentleman here in town that does awesome work. We were basically interviewing him and he picked up a gas tank that was in our shop and started giving us examples of what detail means to him—how he finishes the gas tank underneath, and trails our stripes on the top, and finishes the underside of the tank so it's as beautiful as the top, and the accuracy of stripes. When we do any sort of pure directional stripe, it has to match from the bearing to a tank to a tail section. We're using laser beams to get all of our stripes correct and everything perfect. You just have to know that somebody is as into detail and quality as you are because they really want their work to be looked at up close so it can be completely scrutinized and approved by somebody from ten feet away. The very first paint job we had done for us felt like a gamble because we were letting go of that control.

"The first bike he painted for us was the Turbo Destroyer. Although it was a pretty standard paint job—it was mostly a deep charcoal color with matte and clear-coat and some basic lettering—we could see the quality of his craftsmanship, and that's what we need. It's something you'd see on a show car.

"It's all about detail. You want somebody that is really into the smallest little details.

"We pretty much just provide a sketch now."

A great paint job is worth the time and expense. But there's no denying it's easier to have someone else do the work. *DP Customs, LLC*

CHROME (OR NOT)

After you finish dealing with the paint, there's another material you may want to ditch.

Chrome-plating, if you don't know, is the art of applying chromium onto metal or plastic, giving it the mirror shine of a 1950s-era Chevy bumper. Not so long ago it was the status quo for automotive bling. Chrome rolled down the street on hot rods, choppers, pickup trucks—you name it. Even today, if there's a car show, there will be chrome. And absolutely nothing says it can't work on a motorcycle even now. But judging by our builders, and the custom bikes that appear every day online, chrome is all but dead.

"I like to think of chrome as for industrial use only—fork tubes or master-cylinder pushrods," says John Ryland. "Sometimes header chrome as a rust inhibitor. Anything else always looks better to me matte or brushed. It just got so overused and tacky in the aftermarket that it seems like a substitute for design."

Jarrod DelPrado seconds this: "Justin and I have hated chrome our entire lives. We've always liked stuff that looks more like raw aluminum, or that's powder-coated black, or just something that looks just mean and clean versus trying to look all, 'Hey, look at me—I'm a shiny valve cover or a shiny air cleaner cap.'

"But it's funny—we like polished aluminum, which is probably eighty-five percent close to chrome."

Jared Johnson also prefers to sand down chrome parts using 220-grit sandpaper and then go over it with a Brillo Pad "until it's more of a brushed metal." Even Max Hazan, who tends to rely heavily on raw steel, resulting in bikes that are almost entirely some version of silver or gray, avoids it. "I don't shy away from it because it's shiny, it just looks cheap," he says. "I usually either go polished bare material or nickel plate."

Besides the general distaste for chrome, it can be a stretch to find someone with a good level of skill at plating, let alone considering doing it yourself. Any chrome-plating job worth doing basically requires perfection.

"It's a pretty elaborate and difficult process to do well," says DelPrado. "There are chromers out there who are artists that know how to do it and do it right, versus cheesy, tinted chrome that comes on crappy, cheap motorcycle parts. It's a big art, I think, to do it right, just like powder-coating or nickel-plating or any of that stuff."

Best bet? Either stay away from the stuff or take the time to find someone who can show you samples of superior work—and expect to pay through the nose.

There's something about the raw-steel look that just can't be beaten by outdated, flashy chrome. This 1975 Moto Guzzi is by Revival Cycles.
Alan Stulberg

SEATS AND GRIPS

Even after you've thought about the mirrors and turn signals and foot pegs and even the color of the wheels, there is still a detail that can make or break a bike if handled badly enough: the seat.

If you do any substantial customization to the back half of the frame, you'll likely need to craft your own. This involves first figuring out how it will be shaped and how to mount it. A banana-style seat requires a frame that will extend back far enough to support it. A solo-seater can be shorter, but how will it be shaped? Is it a chopped version of a standard two-seater, or is it more of a guitar-pick shape? Will you bend it out of a piece of sheet metal, or will it be a skateboard deck screwed onto the frame?

The trick, according to Jarrod DelPrado, is to spend the time to make sure everything lines up perfectly. If you're making your own seat pan, it has to not only fit the frame but bolt to the tabs you make for mounting it. If it's even a degree off on one side, people will notice. You'll notice, especially, while riding. Precision matters.

Then there's the matter of covering it.

After all the metal and engine work, it can be a strange change of pace to switch your focus to leather and rivets and staples and foam, the last of which will make for a much nicer ride when you're finally ready to take one.

How you approach the seat depends, as always, on what kind of bike you're creating. A typical café racer will have a seat cowl on the back that hides the battery and other electronics, leaving only a small rectangular area for a seat. If this is what you have, a little foam and a simple piece of leather stretched over it may be all you want. But be wary. Too little design here and you can ruin the hard work you did elsewhere; too much design and it could look ridiculous. It's a fine line, and yet somehow there are a billion possibilities.

Every town has leatherworkers. With enough determination and patience, you can probably find one willing to work with you on a completely custom design, even if he or she has never worked on a motorcycle seat before. Alan Stulberg says his seat-maker had never done one before he walked into the shop.

It might be easier, though, if the person is familiar with the subject.

If Revival's Ducati J63 didn't feature a long, dead-straight foundation line of raw steel with a matching tank and tail, its leather seat wouldn't pop nearly as much as it does. It's a stunner, and that leather will only look better over time as the exhaust pipes age and darken. *Alan Stulberg*

"I'll make my own seat pan," says Jared Johnson, "and fiddle around with that for a little bit. Get all the right angles, so the pan fits the frame. Smooth over the edges. I go with household carpet foam and then do multiple layers—however many I think is going to be comfortable and also look good. I glue down the seat foam and then shape the foam, kind of like a surf board, in a way. I'll round the edges. Make sure it's all nice and straight. Then I'll go to the fabric store with an idea of what color I think is going to look nice on here. I'll buy a yard of fabric and then typically hand it off to Ginger at New Church."

John Ryland and Classified Moto use a similar process. "We use a local upholsterer for our seats," he says. "He's a super-talented guy named Roy Baird in Richmond, Virginia, where we're located. He's an old-school badass biker, but he's really creative and funny as hell. In the past, we've shaped the foam in-house and he sewed the covers. He's picky, though, and lately he prefers to do the whole thing himself. We still make the seat pans."

If you still have some ambition left over from the thousands of other decisions you make during a build, you can do it yourself.

Jared Johnson drew, cut, shaped, upholstered, and installed this custom seat pan for one of his signature bikes.
Jared Johnson

The other detail—one you'll be reminded of literally every time you bend your fingers around the handlebar and head out of the driveway—is what goes under those fingers.

It's such a small detail, but a glaringly wrong choice here can throw off the balance of the whole bike, while a good one can class up the joint better than anything you've done before. On a cohesive, deliberate design, no one will even notice the grips as long as they work with the rest of the bike. But this detail is still an opportunity to accentuate other aspects of the design and to leave a little of yourself in your work.

There are loads of options out there for grips, so it shouldn't be difficult to locate a few options that complement your bike. The key thing is that they do just that. Use the grips as an opportunity to match something else, to create a final point of interest, or even to offset another detail. Did you powder-coat your frame bright pink after all? Matching grips would bring that together like nobody's business.

Several of our builders go with off-the-shelf solutions.

"I do pretty much everything but the stitching myself," says Max Hazan. That fact shines through most on this seat, which he made from scratch by putting his woodworking chops to use. But even when they're upholstered, he says, "I am very particular about the shape of the seat, and I find that most times that I just drop off a seat pan and some fabric, I get an upholstered loaf of bread back. I build my seat frame/pan, shape my own foam–usually medium-density Neoprene; it's waterproof and you can shape it on a belt sander–and make the patterns to sew."
David K. Browning / E3

"Virtually all of our grips and our throttle sleeves come from Biltwell," says Jarrod DelPrado. "They have a bunch of different designs that I think could fit just about any bike, and they're just always really high quality. "They work."

Besides that, he points out, Biltwell offers great service, and grips are one of those rare bolt-on pieces that will actually "bolt on." Slide them onto the bar and you're done. The only caveat is to choose the right size grips for your handlebar—generally 1 inch or 7/8 inch.

Alan Stulberg agrees, but without being so specific about his brand loyalty: "Basic stuff like grips we buy off the shelf, but only the best we can find."

Max Hazan and Jared Johnson are the standouts here. Like everything else, Hazan likes to conjure up something special, so when he buys, he goes with what he loves. "I either make my own out of aluminum or use vintage bicycle grips," he says. "For anything else, my go-to are Renthal full-diamond black road-race grips."

Of all our builders, Jared Johnson's grips stand the best chance of burning themselves into a rider's memory. Why? Because he has trouble finding exactly what he needs. "No one makes good-looking grips—the colors, that is. The grips are great, but I always have trouble matching the seat material to the grips. I just can't find anything with the dark brown. It's usually a throw-up beige. So I figured, yeah, I'm going to start making my own grips."

Biltwellinc.com offers a complete range of grip options, which are widely considered high quality and as stylish as they come.

It might actually be faster to roll your own grips out of handpicked vinyl than to hunt down the perfect pair online or at the local speed shop. This is a set of Jared Johnson's custom grips, which match this bike's custom seat.
Jared Johnson

To do this, he heads to the fabric store again. "Basically," he says, "I use one-inch strips of vinyl, and then I'll glue and wrap them onto the handle bar and make two passes or something. Then glue the vinyl on there."

In other words, he glues the beginning of the strip directly onto the handlebar then wraps it around at a slight angle like an Ace bandage until he gets the grip length and density he needs for the handlebar to stay comfortable through a long afternoon in the mountains. Then he applies glue to the backside at the end and adheres the vinyl strip to itself.

It's nothing particularly technical. "I just use a 3M glue—Super 77—and glue it all on there," he continues. "Then I'll go over with safety wire, which is a thin baling wire. Wherever I end up, I'll wire the grip on there, just the same as motocross bikes."

On the throttle side, he glues and wraps the vinyl strip around a throttle tube rather than the handlebar itself.

There are two major upsides to this. First, vinyl comes in a truckload of colors. If you can't find the one you need at your local fabric store, you can order it online and have it in a couple of days. Second, there's just something about a handmade grip. Maybe it's the perfect color-matching you can get from it. Maybe it's the stylized look of the wrapped vinyl folding over itself from one end to the other. Maybe it's simply the fact that it's vinyl instead of some manufactured, textured rubbery thing that melts in the summer and leaves your hands covered in sticky goo.

Maybe it's knowing that no one else has one exactly like it.

Just like the rest of your bike.

THE MOMENT

And that brings you to the moment.

When you finally finish tightening up the handlebar, lining up the controls just right, making that engine purr, and bolting on the seat that finally came back from the upholsterer, there will be a moment. It's one you'll remember. It's one people around you will remember, because they've watched you hole up in your garage every weekend for an entire winter, endured your long rants about how hard it is to find a good painter in this town and how much you wish you knew more about transmissions. It's a moment that can only happen because of the thousands of other moments you made happen before it. You bought a new ignition. You figured out where it should go. You welded it on yourself. You wired it up. Now it's time to put the key in and light that baby up. With any luck, it'll start on the first try. If not, take comfort in knowing it happens to the best of them. It happens to our builders all the time. You'll sort it out. Deep breaths.

**You know the urge.
It's found you daydreaming more than once.**

When it does start up, you should cancel your plans. You're going to want to ride that thing all day long.

You've done something here that no one will be able to take away from you. You looked over the examples. You studied the styles. You reviewed blog post after blog post to develop a sense of what you like and don't like. You went through a lot of beer just staring at that thing. You made the decisions. You bought parts. You made a few yourself. You scraped and cleaned and spit-shined. Painted. Reassembled. Tuned. It's entirely possible you even named the thing.

Now it's yours. This thing in front of you is an expression of you. Somehow, some way, it is a reflection of things you care about, think, prefer. It's proof that you want a certain kind of life. It's a testament to your beliefs. It's your contribution to the secret handshake we all share when we wave at each other out on the road.

You can go back to your soul-sucking job on Monday. You can grow old and complain about kids today and worry that your gray hair is making you look crotchety before your time. None of it will be able to take away the pride you'll feel when you start that engine and wheel your own creation out of your garage for the first time. And quite possibly every time after that.

When you do, be safe. Throw on a helmet. And some gloves. Come back home so you can take it out again tomorrow. Come back home so you can start another build.

You know the urge.

It's found you daydreaming more than once.

Grab the key for the new ignition you installed. Slide on your helmet. Light up your creation. Go.
David K. Browning / E3

ACKNOWLEDGMENTS

Every book takes a swarm of people, strips them of whatever lives they had before, and turns them into a factory of little book-making worker elves. This one involved more metamorphoses than most. Honestly, it's hard to say I even wrote the thing, because it would not have been conceived without its publisher, made informative without its host of featured builders, perfected without its editor, or turned into a glorious piece of art without its designer. Not to mention all the photographers whose work gives this book its beautiful guts.

One by one, here we go.

John Ryland, you handsome son of a gun, thank you endlessly for your quick responses, your honesty and charm, and your diligence in staying involved despite being buried in media attention over your well-deserved association with *The Walking Dead* and its twin post-apocalyptic zombie motorcycles. If you could draw a straight line or diagnose an engine, you wouldn't be half the inspiration you are to the people reading this book.

Jarrod and Justin DelPrado, enthusiastic gentlemen and fellow rats of the Arizona desert, thank you for your wealth of time and your repeated generosity. You are always happy to offer more words, tell more stories, and share more secrets, and the garage builders of the world will be better off for it. Squeaky clean motorcycles aside, your leap from the corporate airplane into a life of passion is the best mark you could possibly have made.

Max Hazan, artist of conviction and bound for legendary status, thank you for always being the first one to respond to yet another request, for sharing your creativity and talent with the world, and for cryin' out loud, building some of the most stunning bikes mankind has ever seen. Your complete lack of fear is a model for the rest of the world. Here's hoping we all follow your lead.

Jared Johnson, poster child for reckless and awesome, thank you for opening up your shop to my curious eyes, pacing the floor while answering all my questions, sharing your stories of broken backs and handmade VW Beetles, and for refining your tastes so well that your bikes practically come bearing your signature. Your style is brave and unique, and we're honored to have made you part of this book.

Alan Stulberg, elusive and philosophical contributor to the new world motorcycle movement, thank you for so freely offering your insights and photo-shooting talent to this book and to the community at large. Your shop, your show, your work, your bikes–these things all bring the custom scene to a new level, and it would not be what it is without you.

David Browning, photographer extraordinaire and Johnny-on-the-spot, thank you for jumping, flying, and sprinting in to help at every chance, for filling this book with your sharp skills, and for generally carrying the torch of motorcycle madness. Without you, man, this book would be a whole lot thinner and not nearly as pretty.

Geoffrey McCarthy, diligent designer and persistent standard-bearer, thank you for being a man who understands how a motorcycle book should look. Every version you've devised has been so good, it's a minor miracle we chose finals.

Dennis Pernu, guardian of style and voice and all things involving words, thank you for letting me sound like me while somehow finessing every sentence into its best form. I was prepared for an editing fight. Instead, you brought reasoned and measured skill and care. I appreciate that very much.

Lee Klancher, maker of the books and the captain of the process, thank you for pitching this idea over pizza in an Austin brewery and for believing I could be the one to pull it off. It is no small feat running a small press while writing your own book and managing all the others. But Octane always delivers quality, so clearly, you got this.

The photographers, you fantastic artists with your killer eyes for detail and light, thank you for bringing every last one of these bikes to life for our readers and for builders everywhere. Your work is most of what makes us all drool over the idea of crafting our own custom machines. Keep at it. It's working.

Iron & Air magazine for being the first to give me a platform in this motorcycle world. An edit here, a suggestion there, a small story or two, next thing I know, I'm a columnist, working to help shape motorcycle culture for the next gen in one of the finest niche magazines on the planet. I'm holding on to every copy. Thanks for creating something worth holding on to.

Thanks to all of you, to all the readers, to all the men and women who will stand facing a donor bike, wrenches in hand, sparks of ambition and spite in their eyes, ready to make the first twist, the first cut, the first weld. It's just you and the machine now. It's yours to destroy, yours to invent.

Get at it.